Predictions in Time Series Using Regression Models

Springer
New York
Berlin
Heidelberg
Barcelona
Hong Kong
London
Milan
Paris
Singapore
Tokyo

František Štulajter

Predictions in Time Series Using Regression Models

 Springer

František Štulajter
Department of Statistics
Comenius University
FMFI UK
Mlynska Dolina
Bratislava, 842 48
Slovak Republic
frantisek.stulajter@fmph.uniba.sk

Library of Congress Cataloging-in-Publication Data
Štulajter, František.
    Predictions in time series using regression models / František Štulajter.
    p. cm.
    Includes bibliographical references and index.

    1. Time-series analysis.   2. Regression analysis.   I. Title.
QA280 .S82 2002
519.5′5—dc21   2001048431

Printed on acid-free paper.

Photocomposed pages prepared from the author's Scientific Word files.

9 8 7 6 5 4 3 2 1

ISBN 978-1-4419-2965-5

Springer-Verlag   New York Berlin Heidelberg
A member of BertelsmannSpringer Science+Business Media GmbH

To Viera, Katarina, and Andrea

Preface

Books on time series models deal mainly with models based on Box-Jenkins methodology which is generally represented by *autoregressive integrated moving average* models or some nonlinear extensions of these models, such as *generalized autoregressive conditional heteroscedasticity* models. Statistical inference for these models is well developed and commonly used in practical applications, due also to statistical packages containing time series analysis parts.

The present book is based on regression models used for time series. These models are used not only for modeling mean values of observed time series, but also for modeling their covariance functions which are often given parametrically. Thus for a given finite length observation of a time series we can write the regression model in which the mean value vectors depend on regression parameters and the covariance matrices of the observation depend on variance-covariance parameters. Both these dependences can be linear or nonlinear.

The aim of this book is to give an unified approach to the solution of statistical problems for such time series models, and mainly to problems of the estimation of unknown parameters of models and to problems of the prediction of time series modeled by regression models.

The book consist of five chapters. Since many problems of parameter estimation can be formulated in terms of projections, the first chapter starts with the elements of Hilbert spaces and projection theory. In this chapter the basic principles and results on estimation are given, including the method of double least squares estimation. It also contains sections dealing with invariant and unbiased invariant quadratic estimators which can be omitted at the first reading.

In the second chapter the basic models for random processes and time series are described, including the basic results of spectral theory.

The third chapter is devoted to the problems of estimation of regression parameters of time series models. It contains methods and results on the estimation of mean value parameters for both *linear* and *nonlinear regression models* (LRM/NRM). It also contains the *double ordinary least squares estimators* (DOOLSEs) of values of the covariance function which is assumed to be stationary, and sufficient conditions under which these estimators are consistent. An important part of this chapter is devoted to the maximum likelihood method of the estimation of parameters of mean values and covariance functions in the Gaussian case.

In the fourth chapter the problems of the prediction of time series modeled by regression models are studied. It not only contains the classical part on predictions in LRM, but also parts on predictions in cases when multivariate or NRM for mean values are used.

The fifth chapter is devoted to the practical problems of prediction when both the mean value and the covariance function of time series are unknown and are modeled by some reression models. The properties of empirical predictors and some numerical examples are given in this chapter.

The book contains many examples. These examples are often motivated by models from Box-Jenkins methodology, but with mean values modeled by regression models.

I believe that this book will be a useful approach not only for students of mathematical statistics, but also for students and researchers of economical and financial mathematics and management. For these people, as well as for post-graduate students of economic colleges, interested also in the mathematical background of time series modeling, it can serve as complementary to the great amount of literature dealing with the econometric analysis of time series.

František Štulajter
Bratislava
June 21, 2001.

Contents

1

Hilbert Spaces and Statistics

Statistical methods for time series modeled by regression models are based mainly on the theory of Hilbert spaces. We now briefly summarize results which will be used in this book.

1.1 Hilbert Spaces and Projections

Hilbert spaces play an important role in statistics. Many statistical problems of estimation can be formulated as problems of prediction in some Hilbert space. In the sequel, we shall give some special types of simple Hilbert spaces which will often be used in this book.

We shall consider mostly *the Euclidean linear space E^n of $n \times 1$ real vectors* with *an inner product* $(.,.)$ defined by

$$(a, b) = \sum_{i=1}^{n} a_i b_i = a'b,$$

or with an inner product $(.,.)_{\Sigma^{-1}}$ defined by

$$(a, b)_{\Sigma^{-1}} = \sum_{i=1}^{n} \sum_{j=1}^{n} a_i b_j \Sigma_{ij}^{-1} = a' \Sigma^{-1} b$$

for a symmetric *positive definite* (p.d.) $n \times n$ matrix Σ^{-1}. Then $\|a\|^2 = a'a$ and $\|a\|_{\Sigma^{-1}}^2 = a'\Sigma^{-1}a$ is *a squared norm* of a.

The other important Hilbert space is *a space S^n of symmetric $n \times n$ matrices* with an inner product $(.,.)$ defined by

$$
\begin{aligned}
(A, B) &= \sum_{i=1}^{n} \sum_{j=1}^{n} A_{ij} B_{ij} = tr(AB) \\
&= Vec(A)'Vec(B),
\end{aligned}
$$

or with an inner product $(.,.)_{\Sigma^{-1}}$ defined by

$$
\begin{aligned}
(A, B)_{\Sigma^{-1}} &= tr(A\Sigma^{-1}B\Sigma^{-1}) = tr(\Sigma^{-1/2}A\Sigma^{-1/2}\Sigma^{-1/2}B\Sigma^{-1/2}) \\
&= \sum_{i=1}^{n}\sum_{j=1}^{n}\sum_{k=1}^{n}\sum_{l=1}^{n} A_{ik}B_{il}\Sigma_{kj}^{-1}\Sigma_{lj}^{-1} \\
&= Vec(A)'[\Sigma^{-1}\otimes\Sigma^{-1}]Vec(B),
\end{aligned}
$$

where $tr(A) = \sum_{i=1}^{n} A_{ii}$, $Vec(A) = (A'_{.1}A'_{.2}...A_{.n})'$ is the $n^2 \times 1$ vector of columns of A and Σ is a p.d. symmetric matrix and $\otimes$ denotes the Kronecker product.

In E^n we shall consider a *linear space* $L(F) = \{F\beta; \beta \in E^k\}$ generated by columns of an $n \times k$ matrix F and in the space S^n a *linear subspace* K^n of all symmetric $n \times n$ matrices K for which $K_{ij} = K_{|i-j|}; i, j = 1, 2, ..., n$. An $(.,.)$ orthogonal base for K^n are matrices $V^k; k = 0, 1, ..., n-1$ defined by $V_{ij}^k = 1$ if $|i - j| = k$ and $V_{ij}^k = 0$ elsewhere.

For $V = \{V_1, ..., V_l\}$ a set of matrices from S^n, we denote by $L(V)$ *the linear subspace of S^n spanned by the elements of V.*

Let us consider a Hilbert space H with an inner product $(.,.)$. Very important in a Hilbert space is *the Schwarz inequality*

$$|(g, h)| \leq \|g\| \, \|h\| .$$

Using the Schwarz inequality we get

$$
\|AB\|^2 = \sum_{i=1}^{n}\sum_{j=1}^{n}\left(\sum_{k=1}^{n} A_{ik}B_{kj}\right)^2 \leq \sum_{i=1}^{n}\sum_{k=1}^{n} A_{ik}^2 \sum_{l=1}^{n}\sum_{j=1}^{n} B_{lj}^2 = \|A\|^2 \, \|B\|^2 .
$$

Another important notion is a notion of *projection*. Let L be *a linear subspace of H.* Then for any $h \in H$ there exists a unique $h_0 \in L$ such that

$$\|h - h_0\|^2 = \min_{g \in L} \|h - g\|^2$$

and $(h - h_0, g) = 0$ for all $g \in L$. The element h_0 is called an orthogonal, with respect to $(.,.)$, *projection of $h \in H$ on L* and will be denoted by $P_L h$ or Ph. The projection Ph is the unique element in L which has the smallest distance from h.

The element $h - h_0$ is an orthogonal projection on *the orthogonal complement $L^\perp$ to L* defined by

$$L^\perp = \{h \in H : (h, g) = 0 \text{ for all } g \in L\}$$

and will be denoted by $P_{L^\perp} h$, or Mh.

Now let $L = L\{h_1, ..., h_m\}$, then Ph can be written as

$$Ph = \sum_{i=1}^{m} c_i h_i,$$

where

$$Mh = (h - Ph) \perp h_i, i = 1, 2, ..., m,$$

or

$$\left(h - \sum_{i=1}^{m} c_i h_i, h_j\right) = 0; j = 1, 2, ..., m.$$

This can be written as

$$\sum_{i=1}^{m} (h_i, h_j) c_j = (h, h_j); j = 1, 2, ..., m,$$

or, in matrix form,

$$Gc = r,$$

where G is a symmetric $m \times m$ *Gramm matrix of* L given by

$$G_{ij} = (h_i, h_j); i, j = 1, 2, ..., m, c = (c_1, ..., c_m)'$$

and

$$r = ((h, h_1), (h, h_2), ..., (h, h_m))'.$$

From this we get

$$c = G^- r,$$

where G^- is *a generalized inverse* (g-inverse) *of* G given by the equality $GG^- G = G$.

For a symmetric *nonnegative definite* (n.d.) matrix A there exists a generalized inverse A^+ which is also symmetric and for which $A^+ A A^+ = A^+$.

The matrix G^- is p. d. if $h_i; i = 1, 2, ..., m$, are *linearly independent* vectors. In this case, $G^- = G^+ = G^{-1}$ and

$$c = G^{-1} r.$$

Moreover, it is a diagonal matrix if the vectors $h_i; i = 1, 2, ..., m$, are mutually *orthogonal*.

Example 1.1.1. Let us consider E^n with the inner product $(., .)$. Let L be a linear subspace $L(f_1, f_2, ..., f_k)$ spanned by vectors $f_i; i = 1, 2, ..., k$, from E^n. Let us denote by F the $n \times k$ matrix with columns $f_i; i = 1, 2, ..., k$. Then since

$$G_{ij} = (f_i, f_j) = (F'F)_{ij}; i, j = 1, 2, ..., k$$

we have $G = F'F$ and the projection P_L on L can be written using an $n \times n$ symmetric *projection matrix*

$$P = F(F'F)^- F'.$$

For any $x \in E^n$ we can write

$$P_L x = Px = F(F'F)^- F'x.$$

For the matrix P we have $P = P', P^2 = P$, and $r(P) = tr(P)$. There are many g-inverses $(F'F)^-$ of $F'F$, but the projection Px of any vector $x \in E^n$ is unique. The matrix P is a full rank matrix, $r(P) = k$, if the vectors $f_1, ..., f_k$ are linearly independent. In this case we have

$$P = F(F'F)^{-1}F'.$$

The matrix $M = I_n - P$ is *a projection matrix on the subspace* $L(F)^\perp$ with the same properties as the matrix P, but with $r(M) = n - k$ if $r(P) = k$.

If we use the inner product $(.,.)_{\Sigma^{-1}}$, where Σ is a p.d. matrix, then we get the projection matrix P_Σ on $L(F)$ given by

$$P_\Sigma = F(F'\Sigma^{-1}F)^- F'\Sigma^{-1}$$

for which we get

$$P_\Sigma^2 = P_\Sigma, (\Sigma^{-1}P_\Sigma)' = \Sigma^{-1}P_\Sigma, r(P_\Sigma) = k \text{ if } r(F) = k.$$

There are many g-inverses of the Gramm matrix but the projection $P_\Sigma x$ is again unique. The matrix $M_\Sigma = I_n - P_\Sigma$ is a projection matrix on the subspace

$$L(F)^{\perp_\Sigma} = \left\{ x \in E^n : (x, f_i)_{\Sigma^{-1}} = 0; i = 1, 2, ..., k \right\}.$$

It is clear from the definition of a projection on L that $Px = x$ for $x \in L$ and $Px = 0$ for $x \in L^\perp$. By analogy, $Mx = 0$ for $x \in L$ and $Mx = x$ for $x \in L^\perp$.

There are connections between the projection matrices P and P_Σ which are given by the equalities

$$P_\Sigma P = P = P' = PP'_\Sigma$$

from which

$$M_\Sigma = M_\Sigma M, M_\Sigma' = MM_\Sigma',$$

and

$$PP_\Sigma = P_\Sigma$$

which gives

$$M = MM_\Sigma = M_\Sigma' M$$

for any p.d. matrix Σ.

Example 1.1.2. If $(.,.)_{\Sigma^{-1}}$ is an inner product in the space S^n then the projection on a subspace $L(V_1, ..., V_l)$, where $V_1, ..., V_l$ are symmetric linearly independent matrices, is given by a Gramm matrix G with elements

$$G_{ij} = tr(V_i \Sigma^{-1} V_j \Sigma^{-1}); i, j = 1, 2, ..., l.$$

For any matrix $S \in S^n$ we have

$$PS = \sum_{i=1}^{l} \sum_{j=1}^{l} (G^{-1})_{ij} (S, V_j)_{\Sigma^{-1}} V_i,$$

where

$$(S, V_j)_{\Sigma^{-1}} = tr(S\Sigma^{-1}V_j\Sigma^{-1}); j = 1, 2, ..., l.$$

In a special case, when $\Sigma = I_n$, we get

$$\begin{aligned} G_{ij} &= (V_i, V_j) = tr(V_i V_j), \\ (S, V_j) &= tr(SV_j); i, j = 1, 2, ..., l. \end{aligned}$$

If $V_j; j = 1, 2, ..., l$, are *orthogonal matrices* and $(V_i, V_j) = 0$ for $i \neq j$, then G is a *diagonal matrix* with elements $G_{ii} = \|V_i\|^2$ and the matrix G^{-1}

is diagonal with $(G^{-1})_{ii} = \|V_i\|^{-2}; i = 1, 2, ..., l$. In this case

$$PS = \sum_{i=1}^{l} \|V_i\|^{-2} (S, V_i) V_i.$$

Example 1.1.3. An important example of a Hilbert space in statistics is a *Hilbert space* L^2 of random variables Z with $E[Z] = 0$ and a finite variance $D[Z]$. The inner product in this space is defined by

$$(Z, Y)_{L^2} = E[ZY] = Cov(Z; Y).$$

If $X_1, ..., X_n$ are some random variables from L_2, then the Gramm matrix G of $L = L(X_1, ..., X_n)$ is given by

$$G_{ij} = E[X_i X_j] = Cov(X_i; X_j); i, j = 1, 2, ..., n.$$

The projection $P_L U = U^*$ of $U \in L^2$ is given by

$$U^* = \sum_{i=1}^{n} \sum_{j=1}^{n} G_{ij}^{-}(U, X_j)_{L^2} X_i.$$

We can write

$$U^* = \arg \min_{X \in L} E[U - X]^2.$$

If the members of L are linearly independent, then

$$U^* = r'\Sigma^{-1} X,$$

where $X = (X_1, ..., X_n)'$ is a random vector, r is *a vector of covariances* between X and U given by $r = (Cov(X_1; U), ..., Cov(X_n; U))'$, and Σ is *a covariance matrix* of $X, \Sigma = Cov(X)$.

1.2   Preliminaries from Statistics

In the last example of the preceding section we have introduced the notion of a random vector X. Now we shall give the basic properties and models for an *n-dimensional random vector* $X = (X_1, ..., X_n)'$. This is defined as an ordered set of random variables. Realizations of a random vector X will be denoted by x where $x = (x_1, ..., x_n)'$. The space of all possible realizations x of X will be denoted by $\mathcal{X}$ and in statistics this is called a *sample space*. It is a subset of E^n. The *probability distribution* P of X describes fully the random vector X, because the probabilities $P(X = x)$ are defined for all $x \in \mathcal{X}$ if X is *discrete*. The distribution of a *continuous random vector* X is given by its *density* f. Then $P(X \in B) = \int_B f(x)dx_1...dx_n$ for every Borel set B in E^n. The main characteristics of a random vector are *the mean value* $E[X] = \mu$ and *the covariance matrix* $Cov(X) = \Sigma$. The mean value $E[X]$ is called a *characteristic of the location of* X, because the realizations x of X are around $E[X]$.

This is defined by

$$E[X] = (E[X_1], ..., E[X_n])',$$

where $E[X_i]$ denotes the mean value of a random variable X_i.

The *covariance* $Cov(X; Y)$ between the random variables X and Y is defined by

$$Cov(X; Y) = E[(X - E[X])(Y - E[Y])].$$

It holds that $Cov(X; X) = D[X]$ is a *variance of X*. The *covariance matrix* $Cov(X) = \Sigma$ of a random vector X is a symmetric, $n \times n$ n.d. matrix defined by

$$\Sigma_{ij} = Cov(X_i; X_j); i, j = 1, 2, ..., n.$$

We shall write $\Sigma \geq 0$. From the Schwarz inequality we get

$$|\Sigma_{ij}| \leq (\Sigma_{ii}\Sigma_{jj})^{1/2}; i, j = 1, 2, ..., n.$$

The covariance matrix Σ contains information about the *linear dependence* between elements of a random vector.

The mean value and covariance matrix are the main characteristics of a random vector, but they do not describe its distribution. This is given by a *density* if the random vector is continuous. Very important is *the Gaussian, or normal distribution*, given by the density

$$f_{\mu,\Sigma}(x) = \frac{1}{(2\pi)^{n/2} \det(\Sigma)^{1/2}} \exp\left\{-\frac{1}{2}(x - \mu)'\Sigma^{-1}(x - \mu)\right\}; x \in E^n,$$

where $\mu = (\mu_1, ..., \mu_n)' \in E^n$ and $\Sigma \in S^n$, Σ a p.d. matrix, are *parameters of a Gaussian distribution*. It is known that $\mu = E[X]$ and $\Sigma = Cov(X)$. We shall also write $X \sim N_n(\mu, \Sigma)$. Very important are the following well-known properties of a Gaussian distribution.

Let $X \sim N_n(\mu, \Sigma)$, then:

$AX \sim N_m(A\mu, A\Sigma A')$ for any $m \times n$ matrix A of rank m.

$a'X + b \sim N_1(a'\mu + b, a'\Sigma a)$ for any $a \in E^n$ and $b \in E^1$.

If for every $a \in E^n$, $a'X \sim N_1(a'\mu, a'\Sigma a)$ for some $\mu \in E^n$, and some n.d. $\Sigma \in S^n$, then $X \sim N_n(\mu, \Sigma)$.

If $Y = (X_1, ..., X_n, U)' \sim N_{n+1}((\mu, E[U])', \Sigma^Y)$, where

$$\Sigma^Y = Cov(Y) = \begin{pmatrix} \Sigma & r \\ r' & D[U] \end{pmatrix}, r = Cov(X; U), \text{ and } \Sigma = Cov(X),$$

then

$$E[U \mid X] = E[U] + r'\Sigma^{-1}(X - \mu).$$

This is an important property of a Gaussian distribution: the *conditional mean value* $E[U \mid X]$ is *a linear function* of X.

Another property is that the components of a Gaussian random vector X are *independent* iff $Cov(X)$ is a diagonal matrix. For a random vector X with *independent, identically distributed* (i.i.d.) random variables $X_1, ..., X_n$ we have

$$E[X] = \mu j_n \text{ and } Cov(X) = \sigma^2 I_n,$$

where $j_n = (1, ..., 1)', \mu = E[X_i]$ and $\sigma^2 = D[X_i]$ for $i = 1, 2, ..., n$, and I_n is the $n \times n$ *identity matrix*.

Let $X_1, ..., X_q$ be $n \times 1$ random vectors. Then the $n \times q$ matrix X with columns $X_1, ..., X_q$:

$$X = (X_1 ... X_q),$$

will be called a *stochastic matrix* with elements, random variables, $X_{ij}; i = 1, 2, ..., n, j = 1, 2, ..., q$. The matrix $E[X]$, with elements $E[X_{ij}]$, is the *mean value of* X. It is clear that if $Y = (A, X) = tr(AX)$ for some $l \times n$ matrix A, then

$$E[Y] = tr(AE[X]) = (A, E[X]).$$

If X is an $n \times 1$ random vector with a mean value μ and a covariance matrix Σ, then the random matrix $Y = (X - \mu)(X - \mu)'$ has the mean value $E[Y] = Cov(X) = \Sigma$.

Now we shall study the properties of a *quadratic form Q* defined by

$$Q(X) = X'AX = \sum_{i=1}^{n} \sum_{j=1}^{n} A_{ij} X_i X_j,$$

where $A \in S^n$ and X is a random vector with $E[X] = \mu$ and $Cov(X) = \Sigma$. It can be shown that

$$E[Q] = E[X'AX] = tr(A\Sigma) + \mu'A\mu = (A, \Sigma) + \mu'A\mu = (A, \Sigma) + \|\mu\|_A^2.$$

This expression depends only on μ and Σ and is independent of the type of the distribution of X. To compute a variance of Q we need to know the fourth moments of X. If the distribution of X is Gaussian then, see Kubáček (1988),

$$D[Q] = D[X'AX] = 2tr(A\Sigma A\Sigma) + 4\mu'A\Sigma A\mu$$

and

$$Cov(X'AX; X'BX) = 2tr(A\Sigma B\Sigma) + 4\mu' A\Sigma B\mu.$$

We can also write

$$E_{\mu,\Sigma}[X'AX] = (A, \Sigma + \mu\mu')$$

and, if $X \sim N_n(\mu, \Sigma)$, for the variances and covariances

$$
\begin{aligned}
D_{\mu,\Sigma}[X'AX] &= 2\|A\|_\Sigma^2 + 4\|A\mu\|_\Sigma^2 \\
Cov_{\mu,\Sigma}(X'AX; X'BX) &= 2(A, B)_\Sigma + 4(A\mu, B\mu)_\Sigma.
\end{aligned}
$$

In mathematical statistics we often use the following *models for X*, or for μ and Σ. Very important, and the most well known is a *linear regression model (*LRM*)*. The *classical LRM* is given by

$$X = F\beta + \varepsilon; E[\varepsilon] = 0, E[\varepsilon\varepsilon'] = Cov(X) = \sigma^2 I_n,$$

where F is a known $n \times k$ *design matrix*, $\beta \in E^k$ are unknown *regression parameters*, $\varepsilon = (\varepsilon_1, ..., \varepsilon_n)'$ is an unobservable *vector of random errors* and $\sigma^2 \in (0, \infty)$ is an unknown variance of the components of X, the *variance parameter*. We can write

$$E_\beta[X] = F\beta = \sum_{i=1}^{k} \beta_i f_i; \beta \in E^k,$$

where $F = (f_1...f_k)$; f_i denotes the ith column of F. In LRMs the dependence of mean values $m_\beta = E_\beta[X]; \beta \in E^k$, on β is linear, this means we have

$$m_{c_1\beta^1 + c_2\beta^2} = c_1 m_{\beta^1} + c_2 m_{\beta^2}$$

for every $\beta^1, \beta^2 \in E^k$ and for every $c_1, c_2 \in E^1$.

The LRM is called *univariate* if

$$X = f\beta + \varepsilon; E[\varepsilon] = 0, Cov(X) = \sigma^2 I_n; \beta \in E^1, \sigma^2 \in (0, \infty),$$

where f is a given vector from E^n.

If $k = 2$, the LRM is called *bivariate*. In bivariate LRM it is often useful to put $f_1 = j_n = (1, 1, ..., 1)'$. If $k \geq 3$, we call the LRM *multiple*.

In *classical LRMs* we assume that the covariance matrices of X depend on a parameter σ^2 and are given by

$$Cov_{\sigma^2}(X) = \sigma^2 I_n; \sigma^2 \in (0, \infty).$$

If we assume that the covariance matrices $Cov_\nu(X) = \Sigma_\nu$ are given by

$$\Sigma_\nu = \sum_{j=1}^l \nu_j V_j,$$

where $V_1, ..., V_l$ are known symmetric matrices, then the multiple LRM is called *mixed* and $\nu = (\nu_1, ..., \nu_l)'$ is called the *vector of the variance-covariance components* of a mixed LRM. The space Υ of possible values of ν contains all such $\nu \in E^l$ that $\Sigma_\nu = \sum_{j=1}^l \nu_j V_j$ is a n.d. matrix. We write

$$\Upsilon = \left\{ \nu \in E^l : \sum_{j=1}^l \nu_j V_j \geq 0 \right\}$$

In many mixed LRMs we have $V_1 = I_n$ -the identity $n \times n$ matrix. In any mixed LRM the dependence of covariance matrices $\Sigma_\nu; \nu \in \Upsilon$, on the parameter ν is *linear*.

There are also families of covariance matrices $\{\Sigma_\nu; \nu \in \Upsilon\}$ in which the *dependence on parameter ν is nonlinear*. As an example we give

$$\Sigma_{\nu,ij} = \sigma^2 e^{-\alpha|i-j|} \cos \lambda(i - j); i, j = 1, 2, .., n,$$

where $\nu = (\sigma^2, \alpha, \lambda)' \in \Upsilon = (0, \infty) \times (0, \infty) \times \langle -\pi, \pi \rangle$. We see that in this example the elements $\Sigma_{\nu,ij}$ of Σ_ν depend only on $|i - j|; i, j = 1, 2, ..., n$.

Every LRM has the $(k + l) \times 1$ parameters $\theta = (\beta', \nu')'$ belonging to *the parametric space* $\Theta = E^k \times \Upsilon$.

The *multivariate linear regression model (*MLRM*)* is given as a set

$$X_i = F\beta_i + \varepsilon_i; E[\varepsilon_i] = 0; Cov(X_i) = \Sigma_n; i = 1, 2, ..., q,$$

of LRMs. Let X be an $n \times q$ random matrix given by $X = (X_1...X_q)$. Then the MLRM can be written as

$$X = FB + \varepsilon; E[\varepsilon] = 0,$$

where F is a known $n \times k$ design matrix of the model, $B = (\beta_1...\beta_q)$ is a $k \times q$ *matrix of unknown regression parameters,* and $\varepsilon = (\varepsilon_1...\varepsilon_q)$ is an $n \times q$ *matrix of random errors.* We use an MLRM if we observe q random variables on n objects.

Using the operation Vec we can write an MLRM as a multiple one

$$VecX = (I_q \otimes F)VecB + Vec\varepsilon, E[Vec\varepsilon] = 0.$$

For an $nq \times nq$ covariance matrix

$$\Sigma = Cov(VecX) = Cov(X_1', ..., X_q')' = E[(Vec\varepsilon)(Vec\varepsilon)']$$

we can use the model

$$\Sigma = \Sigma_q \otimes \Sigma_n,$$

where Σ_q is a $q \times q$, $\Sigma_n = Cov(X_i); i = 1, 2, ..., q$, is an $n \times n$ covariance matrix, and $\otimes$ denotes the Kronecker product.

In *classical MLRMs*, $\Sigma_n = I_n$. It is possible to use linear or nonlinear parametric models for the covariance matrices Σ_q and Σ_n to avoid problems of overparametrization of an MLRM.

The *replicated linear regression model* (RLRM) is given as a set of independent random vectors

$$X_i = F\beta + \varepsilon_i; E[\varepsilon_i] = 0; Cov(X_i) = \Sigma_n; i = 1, 2, ..., q,$$

of LRMs. Let X be an $n \times q$ random matrix given by $X = (X_1...X_q)$. Then the RLRM can be written as

$$X = F\beta j_q' + \varepsilon; E[\varepsilon] = 0,$$

where F is a known $n \times k$ design matrix of the model, j_q is the $q \times 1$ vector $j_q = (1, 1, ..., 1)'$, and $\varepsilon = (\varepsilon_1...\varepsilon_q)$ is an $n \times q$ *matrix of random errors*. We use an RLRM if we observe q random vectors with the same mean values given by an LRM.

Using the operation Vec we can write an RLRM as a multiple one

$$VecX = (j_q \otimes F)\beta + Vec\varepsilon, E[Vec\varepsilon] = 0, Cov(Vec(X)) = I_q \otimes \Sigma_n.$$

Other parametric models for the mean values of X are $E_\gamma[X] = m_\gamma$; where $\gamma = (\gamma_1, ..., \gamma_q)' \in \Gamma$ and the dependence of m_γ on parameter γ is nonlinear. These models are called *nonlinear regression models* (NRMs) and can be written as

$$X = m_\gamma + \varepsilon; E[\varepsilon] = 0, Cov(X) = \Sigma; \gamma \in \Gamma \subset E^q.$$

In a *classical NRM* we assume that $\Sigma = \sigma^2 I_n; \sigma^2 \in (0, \infty)$, but we can also assume that the covariance matrices Σ depend on some parameter ν and belong to a family $\Xi = \{\Sigma_\nu; \nu \in \Upsilon\}$ which can be the same as in mixed

LRMs, where the dependence on ν is linear, or to a family in which the dependence on ν is nonlinear. Then the parameters of NRMs are

$$\theta = (\gamma', \nu')'; \theta \in \Theta = \Gamma \times \Upsilon,$$

where Θ is the parametric space for this NRM.

As an example we give

$$E_\gamma[X_i] = \beta_1 \cos \lambda i + \beta_2 \sin \lambda i; i = 1, 2, ..., n,$$

where $\gamma = (\beta_1, \beta_2, \lambda)' \in E^1 \times E^1 \times \langle -\pi, \pi \rangle$.

1.3   Estimation of Parameters

In statistics we assume that we have an observation $x = (x_1, ..., x_n)$ of some random vector X, where x belongs to the sample space $\mathcal{X}$ of X. In *parametric statistical problems* we assume that the distribution P_θ of X is unknown and that it belongs to some *parametric family* $P_\Theta = \{P_\theta; \theta \in \Theta\} = \{f_\theta; \theta \in \Theta\}$, where $\Theta \subset E^q$, of possible distributions. The *problem of estimation* consists in estimating the unknown parameter $\theta \in \Theta$, or some *parametric function* $g(\theta); \theta \in \Theta$, with values in E^1 from the data x. *Estimators* should be some functions of X. We shall use the notation $\tilde{\theta}(X), \hat{\theta}(X), \theta^*(X)$ for estimators of θ and $\tilde{g}(X), \hat{g}(X), g^*(X)$ for estimators of g. Estimators are random variables which are dependent on X. The values $\tilde{\theta}(x); x \in \mathcal{X}$, of an estimator $\tilde{\theta}$ are called *estimates*. The quality of an estimator is characterized by its *mean squared error (MSE) function* defined on Θ by

$$MSE_\theta[\tilde{g}(X)] = E_\theta[\tilde{g}(X) - g(\theta)]^2; \theta \in \Theta,$$

or by the MSE *matrix* defined by

$$MSE_\theta[\tilde{\theta}(X)] = E_\theta[(\tilde{\theta}(X) - \theta)(\tilde{\theta}(X) - \theta)']; \theta \in \Theta.$$

We have

$$MSE_\theta[\tilde{g}(X)] = D_\theta[\tilde{g}(X)] + [E_\theta[\tilde{g}(X)] - g(\theta)]^2; \theta \in \Theta.$$

The second term in the last expression is called the *squared bias* of an estimator $\tilde{g}(X)$.

For an *unbiased estimator* $\tilde{\theta}(X)$ for which

$$E_\theta[\tilde{\theta}(X)] = \theta \text{ for all } \theta \in \Theta, \text{ or } E_\theta[\tilde{g}(X)] = g(\theta); \theta \in \Theta,$$

we get

$$MSE_\theta[\tilde{\theta}(X)] = Cov_\theta(\tilde{\theta}(X)) \text{ and } MSE_\theta[\tilde{g}(X)] = D_\theta[\tilde{g}(X)].$$

The quality of an unbiased estimator is described by its covariance matrix. Good estimators minimize the MSE. We call the estimator $g^*(X)$ the *uniformly best unbiased estimator* (UBUE) of g, if it is unbiased, and

$$g^*(X) = \arg\min_{\tilde{g}\in U_g} E_\theta[\tilde{g}(X) - g(\theta)]^2 \text{ for all } \theta \in \Theta,$$

where U_g is a set of all unbiased estimators for g:

$$U_g = \{\tilde{g}(X) : E_\theta[\tilde{g}(X)] = g(\theta) \text{ for all } \theta \in \Theta\}.$$

An estimator $g_0^*(X)$, for which

$$g_0^*(X) = \arg\min_{\tilde{g}\in U_g} E_{\theta_0}[\tilde{g}(X) - g(\theta_0)]^2$$

holds for some $\theta_0 \in \Theta$, is called the *locally, at θ_0, best unbiased estimator* (LBUE) of g. It is clear that g_0^* is the UBUE of g if it does not depend on θ_0.

The well-known *Rao-Cramer inequality* gives the lowest bound of the variance of an unbiased estimator $\tilde{\theta}$. Let the random vector X have a distribution belonging to a family $P_\Theta = \{f_\theta; \theta \in \Theta\}$. Then, under some regularity conditions on P_Θ, the inequality

$$a'Cov_\theta(\tilde{\theta}(X))a \geq a'I(\theta)^{-1}a$$

holds for every $\theta \in \Theta$, for every unbiased estimator $\tilde{\theta}$ of θ and for every $a \in E^q$.

The $q \times q$ matrix $I(\theta)$ defined for every $\theta \in \Theta$ by

$$I(\theta)_{ij} = E_\theta \left[\frac{\partial \ln f_\theta(x)}{\partial \theta_i} \frac{\partial \ln f_\theta(x)}{\partial \theta_j}\right] = E_\theta \left[-\frac{\partial^2 \ln f_\theta(x)}{\partial \theta_i \partial \theta_j}\right]; i,j = 1,2,...,q,$$

is called a *Fisher information matrix*. An estimator $\theta^*(X)$ of θ is called *efficient*, if $Cov_\theta(\theta^*(X)) = I(\theta)^{-1}$ for every $\theta \in \Theta$.

Remark. If the components $X_i; i = 1,2,...,n$ of X are i.i.d., then $I(\theta) = nI_1(\theta)$, where $I_1(\theta)$ is computed using the one-dimensional density $f_\theta(x); x \in E^1$.

The asymptotic properties of estimators are also studied. Let us denote by X^n the vector of observations X, this means let $X^n = (X_1,...,X_n)'$, and let $\tilde{\theta}(X^n)$, or $\tilde{g}(X^n)$, be some estimators.

Then these estimators are called *consistent* if

$$\lim_{n\to\infty} MSE_\theta[\tilde{\theta}(X^n)] = 0 \text{ for all } \theta \in \Theta,$$

or

$$\lim_{n\to\infty} MSE_\theta[\tilde{g}(X^n)] = 0 \text{ for all } \theta \in \Theta.$$

Sufficient conditions for consistency of $\tilde{\theta}$ are

$$\lim_{n\to\infty} D_\theta[\tilde{\theta}(X^n)] = 0 \text{ and } \lim_{n\to\infty} E_\theta[\tilde{\theta}(X^n)] = \theta \text{ for all } \theta \in \Theta.$$

If the last condition is satisfied, the estimator $\tilde{\theta}(X^n)$ is called *asymptotically unbiased*. The same notion is defined by in similar way for an estimator $\tilde{g}(X^n)$ of a parametric function g. Unbiased estimators are consistent if their variances go to zero if n tends to infinity.

Now we describe two basic methods of finding estimators of unknown parameters. To use the first one we have to know the distribution, say the density, $f_\theta(x); x \in E^n$, of X. The function $L_x(.)$, defined on Θ for every $x \in \mathcal{X}$ by

$$L_x(\theta) = f_\theta(x); \theta \in \Theta,$$

is called the *likelihood function* and $\ln L_x(\theta); \theta \in \Theta$, the *loglikelihood function*. An estimator $\tilde{\theta}(X)$ defined for all $x \in \mathcal{X}$ by

$$\tilde{\theta}(x) = \arg\max_{\theta\in\Theta} L_x(\theta) = \arg\max_{\theta\in\Theta} \ln L_x(\theta)$$

is called the *maximum likelihood estimator* (MLE) of θ.

This estimator can be found by solving the *set of likelihood equations*

$$\frac{\partial}{\partial\theta_i} \ln L_x(\theta) \mid_{\theta=\tilde{\theta}(x)} = 0; i = 1, 2, ..., q.$$

In many cases the likelihood equations are nonlinear.

Example 1.3.1. Let $X \sim N_n(\mu j_n, \sigma^2 I); \theta = (\mu, \sigma^2)' \in \Theta = (-\infty, \infty) \times (0, \infty)$. In statistics this random vector X is called a *random sample from a normal distribution*. The loglikelihood function of this random sample is given by

$$\ln L_x(\theta) = -\frac{n}{2}\ln 2\pi - \frac{n}{2}\ln\sigma^2 - \frac{1}{2}\sum_{i=1}^{n}(x_i - \mu)^2; \theta \in \Theta,$$

and it is well known that the MLE $\tilde{\theta}(X) = (\tilde{\mu}(X), \tilde{\sigma}^2(X))'$ is given by

$$\tilde{\mu}(X) = \frac{1}{n} \sum_{i=1}^{n} X_i = \bar{X}, \tilde{\sigma}^2(X) = \frac{1}{n} \sum_{i=1}^{n} (X_i - \bar{X})^2.$$

Moreover, $\bar{X} \sim N_1(\mu, \sigma^2/n)$ and $n/\sigma^2 \tilde{\sigma}^2(X) \sim \chi_{n-1}^2$, $\bar{X}$ is an unbiased estimator of μ and $n/(n-1)\tilde{\sigma}^2(X)$ is an unbiased estimator of σ^2, $\bar{X}$ and $\tilde{\sigma}^2(X)$ are independent, and

$$Cov_\theta(\tilde{\theta}(X)) = \begin{pmatrix} \sigma^2/n & 0 \\ 0 & 2\sigma^4(n-1)/n^2 \end{pmatrix}.$$

It can be seen that the MLEs are consistent in this case. The inverse of the Fisher information matrix is

$$I(\theta)^{-1} = \begin{pmatrix} \sigma^2/n & 0 \\ 0 & 2\sigma^4/n \end{pmatrix}$$

and we see that the MLE $\bar{X}$ is an efficient estimator of μ, the MLE $\tilde{\sigma}^2$ of σ^2 is not efficient, but it is asymptotically efficient.

Example 1.3.2. Let $X \sim N_n(\mu, \Sigma); \mu \in E^n, \Sigma \in S^n$, and let Σ be a p.d. matrix. Then the loglikelihood is given by

$$\ln L_x(\mu, \Sigma) = -\frac{n}{2} \ln(2\pi) - \frac{1}{2} \ln \det(\Sigma) - \frac{1}{2} \|x - \mu\|_{\Sigma^{-1}}^2$$

and the MLE $(\tilde{\mu}, \tilde{\Sigma})$ of (μ, Σ) is given by

$$\begin{aligned}
(\tilde{\mu}(x), \tilde{\Sigma}(x)) &= \arg\max_{\mu,\Sigma} \ln L_x(\mu, \Sigma) \\
&= \arg\max_{\Sigma} \max_{\mu|\Sigma} \ln L_x(\mu, \Sigma) \\
&= \arg\max_{\Sigma} \ln L_x(\tilde{\mu}_\Sigma(x), \Sigma),
\end{aligned}$$

where $\tilde{\mu}_{\Sigma_0}(x) = \arg\max_{\mu} \ln L_x(\mu, \Sigma_0)$ for a given covariance matrix Σ_0. If we have no model on μ and Σ, then $\tilde{\mu}(x) = x$ and $\tilde{\Sigma}(x) = 0$.

In a special case, when we assume that $\mu \in L(F)$ and thus $\mu = F\beta; \beta \in E^k$, we get the loglikelihood function

$$\ln L_x(\beta, \Sigma) = -\frac{n}{2} \ln 2\pi - \frac{1}{2} \ln \det(\Sigma) - \frac{1}{2} \|x - F\beta\|_{\Sigma^{-1}}^2$$

and the MLEs $\tilde{\beta}(x), \tilde{\Sigma}(x)$ for β, Σ are given by

$$\begin{aligned}
(\tilde{\beta}(x), \tilde{\Sigma}(x))' &= \arg\max_{\Sigma} \max_{\beta|\Sigma} \ln L_x(\beta, \Sigma) \\
&= \arg\max_{\Sigma}[-\frac{1}{2} \ln \det(\Sigma) + \max_{\beta|\Sigma}(-\frac{1}{2} \|x - F\beta\|_{\Sigma^{-1}}^2)].
\end{aligned}$$

But for a given Σ we have

$$\arg\max_{\beta|\Sigma}[-\frac{1}{2}\|x - F\beta\|^2_{\Sigma^{-1}}] = \arg\min_{\beta|\Sigma}\|x - F\beta\|^2_{\Sigma^{-1}}$$

and we know from the projection theory that

$$\min_{\beta|\Sigma}\|x - F\beta\|^2_{\Sigma^{-1}} = \left\|x - F\tilde{\beta}_\Sigma(x)\right\|^2_{\Sigma^{-1}},$$

where

$$F\tilde{\beta}_\Sigma(x) = F(F'\Sigma^{-1}F)^{-1}F'\Sigma^{-1}x = P_\Sigma x$$

is the projection of x on $L(F)$ given by the matrix P_Σ. Thus we can write

$$\tilde{\Sigma}(x) = \arg\max_{\Sigma}[-\frac{1}{2}\ln\det(\Sigma) - \frac{1}{2}\left\|x - F\tilde{\beta}_\Sigma(x)\right\|^2_{\Sigma^{-1}}].$$

We see that the loglikelihood function is a nonlinear function of the components $\Sigma_{ij}; i,j = 1,2,...,n$, and thus its maximum should be found by using some numerical iterative methods, and we have no explicit expression for the MLE $\tilde{\beta}(X)$ and $\tilde{\Sigma}(X)$ which are nonlinear functions of X. Moreover, when we have no model on Σ, the model on X is overparametrized, the number $k+n(n+1)/2$ of parameters is larger than n, the number of observations. But also in parametric models on Σ when $\Sigma \in \Xi = \{\Sigma_\nu; \nu \in \Upsilon \subset E^l\}$ we have to solve a nonlinear problem to find the MLE $\hat{\theta} = (\tilde{\beta}', \tilde{\nu}')'$ for $\theta = (\beta', \nu')'$. It is known that in this case the Fisher information matrix $I(\theta)$ is given by

$$I(\theta) = \begin{pmatrix} F'\Sigma_\nu^{-1}F & 0 \\ 0 & \frac{1}{2}tr(\Sigma_\nu^{-1}V_\nu\Sigma_\nu^{-1}V_\nu) \end{pmatrix}$$

where $tr(\Sigma_\nu^{-1}V_\nu\Sigma_\nu^{-1}V_\nu)$ denotes the $l \times l$ matrix with components

$$(\Sigma_\nu^{-1}V_\nu\Sigma_\nu^{-1}V_\nu)_{ij} = tr(\Sigma_\nu^{-1}V_{\nu,i}\Sigma_\nu^{-1}V_{\nu,j}); i,j = 1,2,...,l,$$

and

$$V_{\nu,i} = \frac{\partial\Sigma_\nu}{\partial\nu_i}; i = 1,2,...,l.$$

An exception is if we consider a classical LRM with covariance matrices $\Sigma_{\sigma^2} = \sigma^2 I_n; \sigma^2 \in (0,\infty)$. It is well known that in this case the MLEs for β and σ^2 are

$$\tilde{\beta}(X) = F(F'F)^{-1}F'X$$

and

$$\tilde{\sigma}^2(X) = \frac{1}{n}\sum_{i=1}^{n}(X_i - (F\tilde{\beta})_i)^2.$$

The disadvantage of the maximum likelihood method is that the type of distribution of X should be known. This requirement is often not fulfilled in real situations. Then we can use the following approach originally proposed by Gauss.

Let us consider first a random vector X with i.i.d. components X_i with $E[X_i] = \mu$ and $D[X_i] = \sigma^2; i = 1, 2, ..., n$. Let $\theta = (\mu, \sigma^2)' \in \Theta = (-\infty, \infty) \times (0, \infty)$. Gauss proposed the following method for estimating the unknown mean value μ. He defined the *ordinary least squares estimator* (OLSE) $\hat{\mu}(X)$ for any $x \in \mathcal{X}$ by

$$\hat{\mu}(x) = \arg\min_{\mu}\sum_{i=1}^{n}(x_i - \mu)^2 = \arg\min_{\mu}\|x - \mu j_n\|^2.$$

It can be easily seen, using the direct method of computing, or the projection theory on $L(j_n)$, that

$$\hat{\mu}(x) = \frac{1}{n}\sum_{i=1}^{n}x_i = \bar{x}$$

and we see that the OLSE $\hat{\mu}(X) = \bar{X}$ is the same as the MLE. The advantage of the least squares method is that it is in fact a nonstatistical estimation method which does not depend either on the type of the distribution of X, or on the covariance matrix of X.

This estimator is an unbiased estimator of μ with $D_\theta[\hat{\mu}(X)] = \sigma^2/n$ for any $\theta \in \Theta$ regardless of the type of the distribution of X. Thus the estimator $\bar{X}$ is a consistent estimator of μ.

If we can assume that $E_\mu[X] = \mu j_n; \mu \in (-\infty, \infty)$, and $Cov(X) = \Sigma$, where Σ is a known p.d. matrix, then we can define the *weighted least squares estimator* (WELSE) $\hat{\mu}_\Sigma(X)$ which depends on the covariance matrix Σ of X and is given by

$$\hat{\mu}_\Sigma(x) = \arg\min_{\mu}\|x - \mu j_n\|^2_{\Sigma^{-1}} \; ; x \in \mathcal{X}.$$

Using the projection theory we get

$$\hat{\mu}_\Sigma(X) = (j_n'\Sigma^{-1}j_n)^{-1}j_n'\Sigma^{-1}X = \left(\sum_{i=1}^{n}\sum_{j=1}^{n}\Sigma_{ij}^{-1}\right)^{-1}\sum_{i=1}^{n}\sum_{j=1}^{n}\Sigma_{ij}^{-1}X_j.$$

This is again an unbiased estimator for μ with

$$D_\Sigma[\hat{\mu}_\Sigma(X)] = (j_n'\Sigma^{-1}j_n)^{-1} = \left(\sum_{i=1}^{n}\sum_{j=1}^{n}\Sigma_{ij}^{-1}\right)^{-1}.$$

It is easy to show that for $\Sigma_{\sigma^2} = \sigma^2 I_n; \sigma^2 \in (0,\infty)$, the WELSE $\hat{\mu}_{\sigma^2}$, which will be defined below, does not depend on σ^2 and is identical to both the OLSE and the MLE.

In the special case when $\Sigma = diag(\sigma_i^2)$ we get the WELSE

$$\hat{\mu}_\Sigma(X) = \frac{\sum\limits_{i=1}^{n} \sigma_i^{-2} X_i}{\sum\limits_{i=1}^{n} \sigma_i^{-2}}$$

and

$$D_\Sigma[\hat{\mu}_\Sigma(X)] = \left(\sum_{i=1}^{n}\sigma_i^{-2}\right)^{-1}.$$

This estimator can be consistent under suitable assumptions on variances $\sigma_i^2; i = 1, 2, ..., n$.

We shall assume now that the random vector X follows an LRM

$$X = F\beta + \varepsilon; E[\varepsilon] = 0, E[\varepsilon\varepsilon'] = \Sigma; \beta \in E^k, \Sigma \text{ a p.d. matrix.}$$

The OLSE $\hat{\beta}$ of β is defined by

$$\hat{\beta}(x) = \arg\min_\beta \|x - F\beta\|^2 \,; x \in \mathcal{X},$$

and from the projection theory we get

$$F\hat{\beta}(X) = F(F'F)^{-1}F'X = P_{L(F)}X$$

and thus we have that the OLSE $\hat{\beta}$ of β is given by

$$\hat{\beta}(X) = (F'F)^{-1}F'X.$$

The OLSE is an unbiased estimator for β and the covariance matrix $Cov_\Sigma(\hat{\beta}(X))$ of the OLSE is given by

$$Cov_\Sigma(\hat{\beta}(X)) = (F'F)^{-1}F'\Sigma F(F'F)^{-1}.$$

The WELSE $\hat{\beta}_\Sigma(X)$ for β is defined by

$$\hat{\beta}_\Sigma(x) = \arg\min_\beta \|x - F\beta\|^2_{\Sigma^{-1}} \text{ for every } x \in \mathcal{X},$$

from which we get, using the projection theory,

$$F\hat{\beta}_\Sigma(X) = F(F'\Sigma^{-1}F)^{-1}F'\Sigma^{-1}X = P_\Sigma X$$

and the WELSE $\hat{\beta}_\Sigma$ for β is

$$\hat{\beta}_\Sigma(X) = (F'\Sigma^{-1}F)^{-1}F'\Sigma^{-1}X.$$

This estimator is also called the *Gauss-Markov estimator*, it is unbiased and has a covariance matrix

$$Cov_\Sigma(\hat{\beta}_\Sigma(X)) = (F'\Sigma^{-1}F)^{-1}.$$

It is well known that the WELSE $\hat{\beta}_\Sigma$ is the *best*, at given Σ, *linear unbiased estimator (BLUE)* of β, and thus it can be denoted by β^*_Σ. This estimator also occurs in the formula for the MLEs for a normally distributed X. The problem of the consistency of β^*_Σ will be studied later.

Example 1.3.3. For covariance matrices $\Sigma_{\sigma^2} = \sigma^2 I; \sigma^2 \in (0, \infty)$, we get that the BLUE $\beta^*_{\sigma^2}$ does not depend on σ^2 and in this case we have that the OLSE $\hat{\beta}$ is the *uniformly best linear unbiased estimator (UBLUE)* β^* of β and we can write

$$\beta^*(X) = (F'F)^{-1}F'X$$

and

$$Cov_{\sigma^2}(\beta^*(X)) = \sigma^2(F'F)^{-1}; \sigma^2 \in (0, \infty).$$

This estimator may not be consistent. We shall deal with the problem of the consistency of $\hat{\beta}$ and give sufficient conditions in Chapter 3 dealing with the estimation of parameters of time series.

Example 1.3.4. In an LRM with a design matrix F we can consider the covariance matrices

$$\Sigma_\nu = \sigma_1^2 V_1 + \sigma_2^2 V_2; \nu = (\sigma_1^2, \sigma_2^2)' \in (0, \infty) \times (0, \infty),$$

where

$$V_1 = \begin{pmatrix} I_{n_1} & 0 \\ 0 & 0 \end{pmatrix}, V_2 = \begin{pmatrix} 0 & 0 \\ 0 & I_{n_2} \end{pmatrix}; n_1 + n_2 = n,$$

and I_{n_j} is the $n_j \times n_j$ identity matrix for $j = 1, 2$.

Then for any Σ we have $\Sigma^{-1} = \sigma_1^{-2}V_1 + \sigma_2^{-2}V_2$ and

$$
\begin{aligned}
\hat{\beta}_\Sigma(X) &= \left[F'\left(\sigma_1^{-2}V_1 + \sigma_2^{-2}V_2\right)F\right]^{-1} F'\left(\sigma_1^{-2}V_1 + \sigma_2^{-2}V_2\right) X \\
&= \left[F'\left(V_1 + \frac{\sigma_1^2}{\sigma_2^2}V_2\right)F\right]^{-1} F'\left(V_1 + \frac{\sigma_1^2}{\sigma_2^2}V_2\right) X
\end{aligned}
$$

and we see that the WELSE in this case depends only on the ratio σ_1^2/σ_2^2. In a special case, when $F = j_n$, we get

$$
\begin{aligned}
\hat{\beta}_\Sigma(X) &= \left(n_1 + \frac{\sigma_1^2}{\sigma_2^2}n_2\right)^{-1}\left(\sum_{i=1}^{n_1}X_i + \frac{\sigma_1^2}{\sigma_2^2}\sum_{i=n_1+1}^{n}X_i\right) \\
&= \left(n_1 + n_2\frac{\sigma_1^2}{\sigma_2^2}\right)^{-1}\left(n_1\bar{X}_1 + n_2\frac{\sigma_1^2}{\sigma_2^2}\bar{X}_2\right),
\end{aligned}
$$

where

$$
\bar{X}_1 = \frac{1}{n_1}\sum_{i=1}^{n_1}X_i \text{ and } \bar{X}_2 = \frac{1}{n_2}\sum_{i=n_1+1}^{n}X_i.
$$

This estimator has a variance

$$
D_\Sigma[\hat{\beta}_\Sigma(X)] = \frac{\sigma_1^2\sigma_2^2}{n_1\sigma_2^2 + n_2\sigma_1^2} = \frac{1}{n_1/\sigma_1^2 + n_2/\sigma_2^2}.
$$

Thus $\hat{\beta}_\Sigma$ is, for any given σ_1^2 and σ_2^2, a consistent estimator if $n_1 \to \infty$ and $n_2 \to \infty$. If $\sigma_1^2 = \sigma_2^2 = \sigma^2$, then $\hat{\beta}_\Sigma(X) = \bar{X}$ and $D_\Sigma[\hat{\beta}_\Sigma] = \sigma^2/n$.

Example 1.3.5. Let us consider an LRM with the $n \times 1$ design matrix $F = j_n, n \geq 3$, and with covariance matrices Σ_ν with parameters $\nu = (\sigma^2, \rho)' \in \Upsilon = (0, \infty) \times (-1, 1)$ given by

$$
\Sigma_\nu = \frac{\sigma^2}{1 - \rho^2}\begin{pmatrix}
1 & \rho & \cdots & \rho^{n-2} & \rho^{n-1} \\
\rho & 1 & \cdots & \rho^{n-3} & \rho^{n-2} \\
\cdot & \cdot & \cdots & \cdot & \cdot \\
\rho^{n-2} & \rho^{n-1} & \cdots & 1 & \rho \\
\rho^{n-1} & \rho^{n-2} & \cdots & \rho & 1
\end{pmatrix}.
$$

It is easy to show that, for any $\nu \in \Upsilon$, we can write

$$\Sigma_\nu^{-1} = \frac{1}{\sigma^2} \begin{pmatrix} 1 & -\rho & 0 & \cdots & 0 & 0 & 0 \\ -\rho & 1+\rho^2 & -\rho & \cdots & 0 & 0 & 0 \\ \cdot & & \cdot & \cdots & & & \cdot \\ 0 & 0 & 0 & \cdots & -\rho & 1+\rho^2 & -\rho \\ 0 & 0 & 0 & \cdots & 0 & -\rho & 1 \end{pmatrix}$$

and using these results we get, for the WELSE $\hat{\beta}_\Sigma$, the expression

$$\hat{\beta}_\Sigma(X) = \frac{X_1 + X_n + (1-\rho)\sum\limits_{i=2}^{n-1} X_i}{2 + (n-2)(1-\rho)}$$

and for its variance

$$D_{\sigma^2,\rho}[\hat{\beta}_\Sigma] = (j_n'\Sigma^{-1}j_n)^{-1} = \sigma^2(n - 2(n-1)\rho + (n-2)\rho^2)^{-1}.$$

Thus for any $\rho \in (-1,1)$ we again have a consistent estimator of the expected value β of X_i; $i = 1,2,...,n$. As we have seen the WELSE $\hat{\beta}_\Sigma$ does not depend on σ^2. It depends only on ρ and for $\rho = 0$ we get that the WELSE $\hat{\beta}_\Sigma$ is identical to the OLSE $\bar{X}$. We remark that the dependence of Σ_ν on the parameter ν is nonlinear.

The results just derived can also be used for the estimation of regression parameters in MLRMs as follows. Let

$$VecX = (I_q \otimes F)\beta + Vec\varepsilon;\ E[Vec\varepsilon] = 0;\ Cov(VecX) = \Sigma = \Sigma_q \otimes \Sigma_n,$$

where $\beta = VecB = (\beta_1', ..., \beta_q')'$ are regression parameters, be an MLRM. Let us assume that $\Sigma = \Sigma_q \otimes \Sigma_n$ is a known covariance matrix. Then we can compute the WELSE $\hat{\beta}_\Sigma = (\hat{\beta}_{\Sigma,1}, ..., \hat{\beta}_{\Sigma,q})'$ of β. Using the preceding results for a multiple LRM we get

$$\begin{aligned} \hat{\beta}_\Sigma(X) &= ((I_q \otimes F)'(\Sigma_q \otimes \Sigma_n)^{-1}(I_q \otimes F))^{-1} \\ &\quad (I_q \otimes F)'(\Sigma_q \otimes \Sigma_n)^{-1}(X_1', ..., X_q')' \\ &= (\Sigma_q \otimes (F'\Sigma_n^{-1}F)^{-1}F')(\Sigma_q^{-1} \otimes \Sigma_n^{-1})(X_1', ..., X_q')' \\ &= (I_q \otimes (F'\Sigma_n^{-1}F)^{-1}F'\Sigma_n^{-1})(X_1', ..., X_q')', \end{aligned}$$

from which we get that, independently of Σ_q, the WELSE $\hat{\beta}_\Sigma$ is given by

$$\hat{\beta}_{\Sigma,i}(X) = \hat{\beta}_{\Sigma_n,i}(X_i) = (F'\Sigma_n^{-1}F)^{-1}F'\Sigma_n^{-1}X_i;\ i = 1,2,...,q.$$

If $\Sigma = \Sigma_q \otimes \Sigma_n$, then $Cov(X_i; X_j) = \Sigma_{q,ij}\Sigma_n;\ i,j = 1,2,...,q$, from which

we get

$$Cov(\hat{\beta}_{\Sigma,i}; \hat{\beta}_{\Sigma,j}) = (F'\Sigma_n^{-1}F)^{-1}F'\Sigma_n^{-1}\Sigma_{q,ij}\Sigma_n\Sigma_n^{-1}F(F'\Sigma_n^{-1}F)^{-1}$$
$$= \Sigma_{q,ij}(F'\Sigma_n^{-1}F)^{-1}; i,j = 1, 2, ..., q,$$

or

$$Cov_\Sigma(\hat{\beta}_\Sigma) = \Sigma_q \otimes (F'\Sigma_n^{-1}F)^{-1}.$$

In classical MLRM, where $\Sigma_n = I_n$ and $\Sigma = \Sigma_q \otimes I_n$, we get that the WELSE $\hat{\beta}_\Sigma$ is identical to the OLSE $\hat{\beta} = (\hat{\beta}_1', ..., \hat{\beta}_q')'$, where, for any Σ_q,

$$\hat{\beta}_i(X) = \hat{\beta}_i(X_i) = (F'F)^{-1}F'X_i; i = 1, 2, ..., q,$$

and

$$Cov_\Sigma(\hat{\beta}) = \Sigma_q \otimes (F'F)^{-1}.$$

In an RLRM

$$VecX = (j_q \otimes F)\beta + Vec\varepsilon, E[Vec\varepsilon] = 0, Cov(Vec(X)) = I_q \otimes \Sigma_n,$$

we get the WELSE $\hat{\beta}_\Sigma$ of β in the form

$$\begin{aligned}
\hat{\beta}_\Sigma(X) &= ((j_q \otimes F)'(I_q \otimes \Sigma_n)^{-1}(j_q \otimes F))^{-1}(j_q \otimes F)'(I_q \otimes \Sigma_n)^{-1}Vec(X) \\
&= (j_q'j_q \otimes (F'\Sigma_n^{-1}F))^{-1}(j_q' \otimes F'\Sigma_n^{-1})Vec(X) \\
&= \frac{1}{q}(F'\Sigma_n^{-1}F)^{-1}\sum_{i=1}^{q}F'\Sigma_n^{-1}X_i \\
&= (F'\Sigma_n^{-1}F)^{-1}F'\Sigma_n^{-1}\bar{X},
\end{aligned}$$

where the $n \times 1$ random vector $\bar{X}$ is the arithmetic mean of $X_i; i = 1, 2, ..., q,$

$$\bar{X} = \frac{1}{q}\sum_{i=1}^{q}X_i.$$

Next we have

$$Cov_{\Sigma_n}(\hat{\beta}) = \frac{1}{q}(F'\Sigma_n^{-1}F)^{-1}.$$

The OLSE $\hat{\beta} = (\hat{\beta}_1', ..., \hat{\beta}_q')'$ is given by

$$\hat{\beta}(X) = (F'F)^{-1}F'\bar{X}$$

and

$$Cov_{\Sigma_n}(\hat{\beta}) = \frac{1}{q}(F'F)^{-1}F'\Sigma_n F(F'F)^{-1}.$$

The ordinary and weighted LSE can also be defined for the regression parameters γ of an NRM as follows. The OLSE $\hat{\gamma}$ of γ is given by

$$\hat{\gamma}(x) = \arg\min_{\gamma} \|x - m_\gamma\|^2 = \arg\min_{\gamma} \sum_{i=1}^{n}(x_i - m_{\gamma,i})^2; x \in \mathcal{X},$$

and the WELSE $\hat{\gamma}_\Sigma$, at the given covariance matrix Σ, is given by

$$\hat{\gamma}_\Sigma(x) = \arg\min_{\gamma} \|x - m_\gamma\|^2_{\Sigma^{-1}}; x \in \mathcal{X}.$$

To compute these estimates, which is a nonlinear problem, we must use some iterative method requiring an initial estimator for γ. One such iterative method for computing the OLSE is called *Gauss-Newton* and we shall describe it now.

Let $\hat{\gamma}^{(0)}$ be some initial estimator of γ. Then the iterations for the OLSE are computed according to the expressions

$$\hat{\gamma}^{(i+1)}(x) = \hat{\gamma}^{(i)} + (F'(\hat{\gamma}^{(i)})F(\hat{\gamma}^{(i)}))^{-1}F'(\hat{\gamma}^{(i)})(x - m_{\hat{\gamma}^{(i)}}); i = 0, 1, ...,$$

where $F(\hat{\gamma}) = \partial m_\gamma / \partial\gamma' |_{\hat{\gamma}}$ is an $n \times q$ gradient matrix of m_γ at $\hat{\gamma}$ with elements $F(\hat{\gamma})_{kl} = \partial m_{\gamma,k}/\partial\gamma_l |_{\hat{\gamma}}; k = 1, 2, ..., n, l = 1, 2, ..., q$, and we have used the notation $\hat{\gamma}^{(i)}$ for $\hat{\gamma}^{(i)}(x); i = 1, 2, ...$.

We stop the iterations if

$$\left\|\hat{\gamma}^{(i+1)} - \hat{\gamma}^{(i)}\right\|^2 \le \delta,$$

where δ is a prescribed small number.

A similar iterative procedure is used for computing the WELSE $\hat{\gamma}_\Sigma$ of γ by a given Σ. The $(i+1)$th iteration, $i = 0, 1, ...$ is given by

$$\hat{\gamma}_\Sigma^{(i+1)}(x) = \hat{\gamma}_\Sigma^{(i)}(x) + (F'(\hat{\gamma}_\Sigma^{(i)})\Sigma^{-1}F(\hat{\gamma}_\Sigma^{(i)}))^{-1}F'(\hat{\gamma}_\Sigma^{(i)})\Sigma^{-1}(x - m_{\hat{\gamma}_\Sigma^{(i)}}),$$

where $\hat{\gamma}_\Sigma^{(0)}$ is some initial estimator of γ and the iterations are stopped according to the same rule as for the OLSE.

Since the OLSE $\hat{\gamma}(X)$ and the WELSE $\hat{\gamma}_\Sigma(X)$ are nonlinear functions of X for which we have no explicit expressions, it is difficult to study their properties, mainly for a finite dimension n of X even for i.i.d. components of X. These estimators are, in many NRMs, biased and their biases, as well as their covariance matrices, typically depend on the true value of γ. In the following chapters we shall use some approximations for these estimators to derive their properties including the problem of their consistency.

1.4   Double Least Squares Estimators

Although the least squares method was originally proposed for estimation of regression parameters, it can also be used as a method of estimation of variance-covariance components. By analogy to OLSE and WELSE we can define also variance-covariance estimators of this type. This can be done as follows.

Let $\hat{\beta}$ be the OLSE of β in a classical LRM. Consider the residual vector $\hat{\varepsilon} = (\hat{\varepsilon}_1, ..., \hat{\varepsilon}_n)'$ with components

$$\hat{\varepsilon}_i = X_i - (F\hat{\beta})_i; i = 1, 2, ..., n,$$

where $\hat{\varepsilon}$ is the so-called *vector of ordinary least squares residuals*. This vector can be used for the construction of an estimator S of a covariance matrix $\Sigma = \sigma^2 I_n$ by setting

$$S(X) = (X - F\hat{\beta})(X - F\hat{\beta})' = \hat{\varepsilon}\hat{\varepsilon}'.$$

This estimator, the random $n \times n$ matrix $S(X)$, does not fulfil our model on covariance matrices, but it can be used for estimating the unknown parameter σ^2 of Σ by the so-called *double least squares principle*. According to this principle we define $\hat{\sigma}^2$ by

$$\hat{\sigma}^2(X) = \arg\min_{\sigma} \left\| S(X) - \sigma^2 I_n \right\|^2 = \arg\min_{\sigma} \sum_{i=1}^{n} \sum_{j=1}^{n} (S(X)_{ij} - \sigma^2 I_{n,ij})^2.$$

This estimator is given by

$$\hat{\sigma}^2(X) = \frac{(S(X), I_n)}{\|I_n\|^2} = \frac{1}{n} \sum_{i=1}^{n} (X_i - (F\hat{\beta})_i)^2$$

and we can see that this estimator is equal to the MLE if X has normal distribution. It is not unbiased and has the MSE

$$MSE_{\sigma^2}[\hat{\sigma}^2(X)] = \frac{2n - 2k + k^2}{n^2}\sigma^4; \sigma^2 \in (0, \infty).$$

We shall use the name *double ordinary least squares estimator* (DOOLSE) for this estimator of σ^2.

Example 1.4.1. For a classical univariate LRM with i.i.d. components of X, that is, with $F = j_n$, we get

$$\hat{\sigma}^2(X) = \frac{1}{n} \sum_{i=1}^{n} (X_i - \bar{X})^2$$

and

$$MSE_{\sigma^2}[\hat{\sigma}^2(X)] = \frac{2n - 1}{n^2}\sigma^4.$$

This approach can also be used for a mixed LRM with covariance matrices

$$\Sigma_\nu = \sum_{j=1}^{l} \nu_j V_j; \nu \in \Upsilon.$$

The DOOLSE $\hat{\nu} = (\hat{\nu}_1, ..., \hat{\nu}_l)'$ for ν in a mixed LRM is defined by

$$\hat{\nu}(X) = \arg\min_\nu \sum_{i=1}^{n}\sum_{j=1}^{n} [S(X)_{ij} - \Sigma_{\nu,ij}]^2 = \arg\min_\nu \left\| S(X) - \sum_{j=1}^{l} \nu_j V_j \right\|^2,$$

where the matrix $S(X)$ was already defined. From the projection theory we get

$$\hat{\nu}_j(X) = \sum_{k=1}^{l} G_{jk}^-(S(X), V_k) = \sum_{k=1}^{l} G_{jk}^- tr(S(X)V_k); j = 1, 2, ..., l,$$

where G is the Gramm matrix of the $L(\mathcal{V})$, where $\mathcal{V} = (V_1, .., V_l)$ with elements

$$G_{jk} = (V_j, V_k) = tr(V_j V_k); j, k = 1, 2, ..., l.$$

If $g \in E^l$, then $\tilde{g}(X) = g'\tilde{\nu}(X)$, where $\tilde{\nu}(X) = (\tilde{\nu}_1(X), ..., \tilde{\nu}_l(X))'$, will be called the DOOLSE of the linear parametric function $g(\nu) = g'\nu; \nu \in \Upsilon$. If the matrices $V_j; j = 1, 2, ..., l$, are orthogonal, then

$$\hat{\nu}_j(X) = \frac{(S(X), V_j)}{\|V_j\|^2} = \frac{tr(S(X)V_j)}{tr(V_j^2)}; j = 1, 2, ..., l.$$

Example 1.4.2. Let us consider an MLRM with parametric covariance matrices $\Sigma_\nu = \sigma_1^2 V_1 + \sigma_2^2 V_2; \nu = (\sigma_1^2, \sigma_2^2)' \in (0, \infty) \times (0, \infty)$, where

$$V_1 = \begin{pmatrix} I_{n_1} & 0 \\ 0 & 0 \end{pmatrix} \text{ and } V_2 = \begin{pmatrix} 0 & 0 \\ 0 & I_{n_2} \end{pmatrix}; n_1 + n_2 = n,$$

are orthogonal matrices with $\|V_1\|^2 = n_1$ and $\|V_2\|^2 = n_2$. Then

$$\hat{\sigma}_1^2(X) = \frac{1}{n_1} \sum_{i=1}^{n_1} (X_i - (F\hat{\beta})_i)^2$$

and

$$\hat{\sigma}_2^2(X) = \frac{1}{n_2} \sum_{i=n_1+1}^{n} (X_i - (F\hat{\beta})_i)^2$$

are the DOOLSEs for σ_1^2, σ_2^2. Both these estimators are consistent if n_1 and n_2 tend to infinity.

Example 1.4.3. For time series, as we shall see later, the following families of covariance matrices are important:

$$\Sigma_\nu = \nu_0 I_n + \sum_{j=1}^{n-1} \nu_j V_j; \nu \in \Upsilon,$$

where

$$V_j = \begin{pmatrix} 0 & I_{n-j} \\ 0 & 0 \end{pmatrix} + \begin{pmatrix} 0 & 0 \\ I_{n-j} & 0 \end{pmatrix}; j = 1, 2, ..., n-1.$$

The $n \times n$ matrices $V_j; j = 0, 1, ..., n-1$, are again orthogonal, $\|I_n\|^2 = n$ and $\|V_j\|^2 = 2(n-j); j = 1, 2, ..., n-1$. Next we have

$$tr(S(X)I_n) = \sum_{i=1}^{n}(X_i - (F\hat{\beta})_i)^2$$

and

$$tr(S(X)V_j) = 2\sum_{i=1}^{n-j}(X_{i+j} - (F\hat{\beta})_{i+j})(X_i - (F\hat{\beta})_i); j = 1, 2, ..., n-1.$$

Thus for the DOOLSE $\hat{\nu} = (\hat{\nu}_0, ..., \hat{\nu}_{n-1})'$ we get the expressions

$$\hat{\nu}_j(X) = \frac{1}{n-j}\sum_{i=1}^{n-j}(X_{i+j} - (F\hat{\beta})_{i+j})(X_i - (F\hat{\beta})_i); j = 0, 1, ..., n-1.$$

The problem of the consistency of these estimators will be studied later. For the matrix $S(X)$ we can use the following expressions

$$S(X) = (X - F\hat{\beta})(X - F\hat{\beta})' = (X - PX)(X - PX)' = MX(MX)' = \hat{\varepsilon}\hat{\varepsilon}'$$

and, using these equalities, we can write

$$\begin{aligned} (S(X), V_k) &= tr(S(X)V_k) = tr(MXX'MV_k) \\ &= X'MV_kMX = \hat{\varepsilon}'V_k\hat{\varepsilon}; k = 1, 2, ..., l. \end{aligned}$$

We can see that the DOOLSEs $\hat{\nu}_j; j = 1, 2, ..., l$, are functions of statistics $\hat{\varepsilon}'V_k\hat{\varepsilon}; k = 1, 2, ..., l$, which are quadratic forms in X.

As an estimator of an unknown covariance matrix Σ_ν we can then use the DOOLSE $\hat{\Sigma}_\nu$ defined by

$$\hat{\Sigma}_\nu = \Sigma_{\hat{\nu}} = \sum_{j=1}^{l} \hat{\nu}_j V_j.$$

In a similar way as for regression parameters we can also define the *double weighted least squares estimator (DOWELSE)* for the variance-covariance parameters ν of LRMs. Let Σ be a given covariance matrix from the set $\Sigma_\nu; \nu \in \Upsilon$, of the covariance matrices of an LRM, let $\hat{\beta}_\Sigma$ be the WELSE of β by the given Σ, and let $\hat{\varepsilon}_\Sigma = X - F\hat{\beta}_\Sigma$ be the *vector of weighted least squares residuals*. Let

$$S_\Sigma(X) = (X - F\hat{\beta}_\Sigma)(X - F\hat{\beta}_\Sigma)'.$$

Then the DOWELSE $\hat{\nu}_\Sigma$ for a variance-covariance parameter ν of covariance matrices $\Sigma_\nu; \nu \in \Upsilon$, is defined by

$$\hat{\nu}_\Sigma(x) = \arg\min_\nu \|S_\Sigma(x) - \Sigma_\nu\|_{\Sigma^{-1}}^2 ; x \in \mathcal{X}.$$

For a mixed LRM with $\nu = (\nu_1, ..., \nu_l)'$ and $\Sigma_\nu = \sum_{j=1}^{l} \nu_j V_j; \nu \in \Upsilon$, we get, using the expressions for DOOLSE with the inner product $(.,.)_{\Sigma^{-1}}$ instead of $(.,.)$, that the DOWELSEs $\hat{\nu}_{\Sigma,j}; j = 1, 2, ..., l$ of ν, are given by

$$\hat{\nu}_{\Sigma,j}(X) = \sum_{k=1}^{l} (G_\Sigma^-)_{jk}(S_\Sigma(X), V_k)_{\Sigma^{-1}},$$

where

$$G_{\Sigma,jk} = (V_j, V_k)_{\Sigma^{-1}} = tr(V_j \Sigma^{-1} V_k \Sigma^{-1}); j, k = 1, 2, ..., l,$$

and

$$(S_\Sigma(X), V_k)_{\Sigma^{-1}} = tr(S_\Sigma(X)\Sigma^{-1} V_k \Sigma^{-1}); k = 1, 2, ..., l.$$

The matrix $\hat{\Sigma}_{\nu,\Sigma}$, defined by

$$\hat{\Sigma}_{\nu,\Sigma}(X) = \sum_{j=1}^{l} \hat{\nu}_{\Sigma,j}(X) V_j,$$

is the *DOWELSE of a covariance matrix Σ_ν by the given Σ.*

Example 1.4.4. Let us consider first a classical LRM with a design matrix F. Then we have the covariance matrices $\Sigma_\sigma = \sigma^2 I_n; \sigma^2 \in (0, \infty)$, and, for any $\sigma^2 \in (0, \infty)$, we have $\Sigma^{-1} = \sigma^{-2} I_n$. Thus we get $\hat{\beta}_\Sigma = \hat{\beta}$,

$$(S(X), I_n)_{\Sigma^{-1}} = \sigma^{-4}(S(X), I_n) \text{ and } \|I_n\|_{\Sigma^{-1}}^2 = \sigma^{-4} \|I_n\|^2 = \sigma^{-4} n.$$

From this it is easy to see that the DOWELSE $\hat{\sigma}_\Sigma^2$ does not depend on Σ, or on σ^2, and is equal to the DOOLSE:

$$\hat{\sigma}_\Sigma^2(X) = \hat{\sigma}^2(X) = \frac{1}{n} \sum_{i=1}^n (X_i - (F\hat{\beta})_i)^2$$

for all $\Sigma_\sigma = \sigma^2 I_n; \sigma^2 \in (0, \infty)$.

Example 1.4.5. Let us again consider the mixed LRM studied in Example 1.3.4. Then, for any symmetric $n \times n$ matrix A, we can write

$$(A, V_j)_{\Sigma^{-1}} = tr(A(\sigma_1^{-2} V_1 + \sigma_2^{-2} V_2) V_j (\sigma_1^{-2} V_1 + \sigma_2^{-2} V_2))$$

and, using the equalities $V_j V_j = V_j; j = 1, 2$, and $V_1 V_2 = 0$, we get

$$(A, V_j)_{\Sigma^{-1}} = \sigma_j^{-4} tr(A V_j) = \sigma_j^{-4}(A, V_j); j = 1, 2.$$

Next we get

$$\|V_j\|_{\Sigma^{-1}}^2 = \sigma_j^{-4} \|V_j\|^2 = n_j \sigma_j^{-4}; j = 1, 2, \text{ and } (V_1, V_2)_{\Sigma^{-1}} = 0.$$

Since the matrices V_1 and V_2 are not only $(.,.)$, but also $(.,.)_{\Sigma^{-1}}$ orthogonal, we can write, for the DOWELSE,

$$\hat{\sigma}_{\Sigma,j}^2(X) = \frac{(S(X), V_j)_{\Sigma^{-1}}}{\|V_j\|_{\Sigma^{-1}}^2} = \frac{(S(X), V_j)}{n_j}; j = 1, 2.$$

From these expressions we get

$$\hat{\sigma}_{\Sigma,1}^2(X) = \frac{1}{n_1} \sum_{i=1}^{n_1} (X_i - (F\hat{\beta}_\Sigma)_i)^2,$$

$$\hat{\sigma}_{\Sigma,2}^2(X) = \frac{1}{n_2} \sum_{i=n_1+1}^{n} (X_i - (F\hat{\beta}_\Sigma)_i)^2.$$

The WELSE $\hat{\beta}_\Sigma$ was computed in Example 1.3.4, where it was shown that $\hat{\beta}_\Sigma$ depends only on the ratio σ_1^2 / σ_2^2 and thus also the DOWELSEs $\hat{\sigma}_{\Sigma,j}^2; j = 1, 2$, depend only on this ratio. Moreover, as can be seen from Example 1.3.7., the expressions for the DOWELSEs are the same as for the DOOLSEs, only the *weighted least squares residuals* are used instead of the ordinary least squares residuals.

As before we can write

$$S_\Sigma(X) = (I - P_\Sigma X)(I - P_\Sigma X)' = M_\Sigma X(M_\Sigma X)' = \hat\varepsilon_\Sigma \hat\varepsilon_\Sigma'$$

and

$$
\begin{aligned}
(S_\Sigma(X), V_k)_{\Sigma^{-1}} &= tr(M_\Sigma XX' M_\Sigma' \Sigma^{-1} V_k \Sigma^{-1}) = X' M_\Sigma' \Sigma^{-1} V_k \Sigma^{-1} M_\Sigma X \\
&= \hat\varepsilon_\Sigma' \Sigma^{-1} V_k \Sigma^{-1} \hat\varepsilon_\Sigma.
\end{aligned}
$$

We see that the DOWELSEs $\hat\nu_{\Sigma,j}; j = 1, 2, ..., l$ are again quadratic forms in X.

As we shall see later, the DOWELSE $\hat\nu_\Sigma$ plays an important role in computing the MLE $\tilde\nu$ in mixed LRMs.

The DOWELSE can also be defined for an NRM on mean values $m_\gamma; \gamma \in \Gamma$, of X. Let $\hat\gamma_\Sigma$ be the WELSE for γ at the given Σ and let

$$\hat\varepsilon_\Sigma = X - \hat m_\Sigma \text{ where } \hat m_\Sigma = m_{\hat\gamma_\Sigma}$$

be the vector of the weighted least squares residuals. Then the random $n \times n$ matrix

$$S_\Sigma(X) = (X - \hat m_\Sigma)(X - \hat m_\Sigma)'$$

is an initial estimator of $Cov(X)$. If the covariance matrices of X are modeled by a family $\Sigma_\nu; \nu \in \Upsilon$, where the dependence on the parameter ν can be either linear as in a mixed LRM, or nonlinear, then the DOWELSE $\hat\nu_\Sigma$, at the given Σ, is again defined by

$$\hat\nu_\Sigma(X) = \arg\min_\nu \|S_\Sigma(X) - \Sigma_\nu\|^2_{\Sigma^{-1}}.$$

A computation of the WELSE $\hat\nu_\Sigma$ for the case $\Sigma_\nu = \sum_{j=1}^{l} \nu_j V_j; \nu \in \Upsilon$, can be performed by using the expressions derived for $\hat\nu_{\Sigma,j}; j = 1, 2, ..., l$, for a mixed LRM.

If the dependence of Σ_ν on the parameter ν is nonlinear, then we have to solve a nonlinear minimization problem and we must use some iterative method for computing the DOWELSE $\hat\nu_\Sigma$. We remark that for $\Sigma = \sigma^2 I_n$ the DOWELSE $\hat\nu_\Sigma$ is identical with the DOOLSE $\hat\nu$ defined by

$$\hat\nu(X) = \arg\min_\nu \|S(X) - \Sigma_\nu\|^2,$$

where

$$S(X) = (X - m_{\hat\gamma})(X - m_{\hat\gamma})'$$

and $\hat\gamma$ is the OLSE of γ.

1.5   Invariant Quadratic Estimators

The notion of a quadratic form was introduced in Section 1.3 where expressions for the mean values, variances, and covariances of quadratic forms were given. In LRM, when we assume that $E_\beta[X] = F\beta; \beta \in E^k$, we get from these expressions the equality

$$E_{\beta,\Sigma}[X'AX] = tr(A\Sigma) + \beta'F'AF\beta$$

and, for a normally distributed random vector X with covariance matrix Σ:

$$
\begin{aligned}
D_{\beta,\Sigma}[X'AX] &= 2tr(A\Sigma A\Sigma) + 4\beta'F'A\Sigma AF\beta, \\
Cov_{\beta,\Sigma}(X'AX; X'BX) &= 2tr(A\Sigma B\Sigma) + 4\beta'F'A\Sigma BF\beta.
\end{aligned}
$$

Generally these characteristics of X depend on both β and Σ. In LRMs the notion of an invariant, with respect to a regression parameter β, quadratic form is very important. A quadratic form $Q(X) = X'AX$ is called *invariant* if $AF = 0$, where F is a design matrix of an LRM. Since in an LRM $X = F\beta + \varepsilon$, an invariant quadratic form Q can be written as

$$Q(X) = (F\beta + \varepsilon)'A(F\beta + \varepsilon) = \varepsilon'A\varepsilon \text{ for any } \beta \in E^k,$$

from which we have that invariant quadratic forms do not depend on the regression parameter β. As a consequence, we also get that mean values, variances, and covariances of invariant quadratic forms do not depend on β. This is an important property which can be used by estimating the variance-covariance components of covariance matrices $Cov(X)$ of X. For the invariant quadratic form determined by a symmetric matrix A such that $AF = 0$ we get

$$E_{\beta,\Sigma}[X'AX] = E_\Sigma[X'AX] = (A, \Sigma) = tr(A\Sigma)$$

and, if $X \sim N_n(F\beta, \Sigma)$,

$$D_{\beta,\Sigma}[X'AX] = D_\Sigma[X'AX] = 2tr(A\Sigma A\Sigma)$$

and

$$Cov_{\beta,\Sigma}(X'AX; X'BX) = Cov_\Sigma(X'AX; X'BX) = 2tr(A\Sigma B\Sigma)$$

for every $\beta \in E^k$.

From the expression for the mean value of any invariant quadratic form it can be seen that such a quadratic form can be used as an unbiased estimator of a linear parametric function

$$g(\Sigma) = (A, \Sigma) = tr(A\Sigma) = \sum_{i=1}^{n}\sum_{j=1}^{n} A_{ij}\Sigma_{ij},$$

where Σ are possible covariance matrices of X and A is a given symmetric matrix.

It is well known that in the case when Σ can be any n.d. matrix, the estimator $\hat{g}(X) = X'AX$ is the unique *invariant quadratic unbiased estimator* of $g(\Sigma) = (A, \Sigma); \Sigma \geq 0$.

We remark that the condition $AF = 0$ is a sufficient condition for invariancy. The necessary condition for the independence of an expected value of a quadratic form $X'AX$ on β is $F'AF = 0$. But in this case the variance of $X'AX$ can depend on β. The following example describes such a situation.

Example 1.5.1. Let us consider an LRM with $F = j_n$ and let quadratic forms $Q_t(X); t = 1, 2, ..., n - 1$, be defined by

$$Q_t(X) = \frac{1}{n-t} \sum_{s=1}^{n-t} X_{s+t}X_s - \bar{X}^2,$$

where $\bar{X} = 1/n \sum_{i=1}^{n} X_i$. $Q_t; t = 1, 2, ..., n - 1$, are quadratic forms which can be written as

$$Q_t(X) = \frac{1}{n-t} \sum_{s=1}^{n-t} X_{s+t}X_s - \frac{1}{n^2} \sum_{i=1}^{n}\sum_{j=1}^{n} X_iX_j = X'A_tX$$

with

$$A_t = \frac{1}{2(n-t)}\left[\begin{pmatrix} 0 & I_{n-t} \\ 0 & 0 \end{pmatrix} + \begin{pmatrix} 0 & 0 \\ I_{n-t} & 0 \end{pmatrix}\right] - \frac{1}{n^2}J_n,$$

where I_{n-t} is the $(n-t) \times (n-t)$ identity matrix and $J_n = j_n j'_n$ is the $n \times n$ matrix with all elements equal to 1. Using these expressions for matrices $A_t; t = 1, 2, .., n - 1$ we get that

$$
\begin{aligned}
A_tF &= A_t j_n = \frac{1}{2(n-t)}\left[\begin{pmatrix} 0 & j_{n-t} \\ 0 & 0 \end{pmatrix} + \begin{pmatrix} 0 & 0 \\ j_{n-t} & 0 \end{pmatrix}\right] - \frac{1}{n^2}j_n j'_n j_n \\
&= \frac{1}{2(n-t)}\left[\begin{pmatrix} j_{n-t} \\ 0 \end{pmatrix} + \begin{pmatrix} 0 \\ j_{n-t} \end{pmatrix}\right] - \frac{1}{n}j_n,
\end{aligned}
$$

where j_{n-t} is the $(n - t) \times 1$ vector of ones. Thus $A_tF \neq 0$ for every t and the quadratic forms Q_t are not invariant. But it can easily be easily that

$$F'AF = \frac{1}{2(n-t)}2(n - t) - \frac{1}{n}n = 0$$

and thus the quadratic forms $Q_t(X)$ are unbiased estimators of the linear parametric functions

$$g_t(\Sigma) = (A_t, \Sigma); \Sigma \geq 0, t = 1, 2, ..., n - 1.$$

The variances of these estimators generally depend on parameter β which is an undesirable property.

Example 1.5.2. Let us again consider the LRM from Example 1.4.1 and let us now study the quadratic forms U_t defined by

$$U_t(X) = \frac{1}{n - t} \sum_{s=1}^{n-t} X_{s+t} X_s - \bar{X}_{1,t} \bar{X}_{2,t},$$

where

$$\bar{X}_{1,t} = \frac{1}{n - t} \sum_{i=1}^{n-t} X_i \text{ and } \bar{X}_{2,t} = \frac{i}{n - t} \sum_{i=t+1}^{n} X_i; t = 1, 2, .., n - 1.$$

Since both components of U_t, similar to the preceding example, can be written as quadratic forms, we get

$$U_t(X) = X' B_t X,$$

where

$$B_t = \frac{1}{2(n-t)} \left[\begin{pmatrix} 0 & I_{n-t} \\ 0 & 0 \end{pmatrix} + \begin{pmatrix} 0 & 0 \\ I_{n-t} & 0 \end{pmatrix} \right]$$
$$- \frac{1}{2(n-t)^2} \left[\begin{pmatrix} 0 & J_{n-t} \\ 0 & 0 \end{pmatrix} + \begin{pmatrix} 0 & 0 \\ J_{n-t} & 0 \end{pmatrix} \right];$$

$t = 1, 2, ..., n - 1$. Here J_{n-t} denotes the $(n - t) \times (n - t)$ matrix with all elements equal to 1. Next we have

$$B_t F = \frac{1}{2(n-t)} \left[\begin{pmatrix} j_{n-t} \\ 0 \end{pmatrix} + \begin{pmatrix} 0 \\ j_{n-t} \end{pmatrix} \right]$$
$$- \frac{1}{2(n-t)^2} \left[(n-t) \begin{pmatrix} j_{n-t} \\ 0 \end{pmatrix} + (n-t) \begin{pmatrix} 0 \\ j_{n-t} \end{pmatrix} \right]$$

and we see that $B_t F = 0$ for every t, from which we get that the quadratic forms $U_t; t = 1, 2, ..., n - 1$, are invariant.

For $t = 0$, we get

$$Q_0(X) = U_0(X) = \frac{1}{n} \sum_{i=1}^{n} (X_i - \bar{X})^2$$

which is an invariant quadratic form, since

$$A_0 = B_0 = \frac{1}{n}I_n - \frac{1}{n^2}J_n$$

and

$$A_0 F = A_0 j_n = \frac{1}{n}j_n - \frac{1}{n^2}J_n j_n = \frac{1}{n}j_n - \frac{1}{n}j_n = 0.$$

We remark that the invariant quadratic form Q_0 is equal to the MLE $\tilde{\sigma}^2$ derived in Example 1.3.1, or the DOOLSE $\hat{\sigma}^2$ derived in Example 1.4.4, if we set $F = j_n$. Thus the estimators considered in these examples are invariant quadratic unbiased estimators of σ^2.

A description of invariant quadratic estimators can be given as follows. Let

$$\mathcal{I} = \{A \in S^n : AF = 0\}.$$

Let $M = I_n - P = I_n - F(F'F)^{-1}F$ and let

$$\mathcal{I}_M = \{MBM; B \in S^n\}.$$

It is easy to show, using the equality $M = M^2$, that

$$\mathcal{I}_M = \{A \in S^n : A = MAM\}.$$

Next we have $\mathcal{I}_M \subset \mathcal{I}$, and $AF = 0, A \in S^n$ implies $L(A') = L(A) \subset L(F)^\perp$ and thus $MA = A = A' = AM$, or $A = MAM$. From this we can deduce that

$$\begin{aligned} \mathcal{I} &= \{A \in S^n : AF = 0\} \\ &= \mathcal{I}_M = \{A \in S^n : A = MAM\}. \end{aligned}$$

The random variable MX is called *maximal invariant*. From the last equality we get that every invariant quadratic form $Q(X) = X'AX; A \in \mathcal{I}$, can be written in the form

$$X'AX = X'M'AMX = X'MAMX,$$

or

$$X'AX = (X - F\hat{\beta})'A(X - F\hat{\beta}) = \hat{\varepsilon}'A\hat{\varepsilon},$$

where $\hat{\varepsilon} = X - F\hat{\beta}$ are the ordinary least squares residuals.

The following characterization of $\mathcal{I}$ is also possible. Let

$$\mathcal{N} = \{N : N^2 = N, NF = 0\}$$

be a set of $n \times n$, not necessarily symmetric, matrices and let

$$\mathcal{I}_N = \{B \in S^n : B = N'AN; A \in S^n\}; N \in \mathcal{N}.$$

It is easy to show that

$$\mathcal{I}_N = \{A \in S^n : A = N'AN\}$$

and that $\mathcal{I}_N \subset \mathcal{I}$ for every $N \in \mathcal{N}$ and thus $\bigcup_{N \in \mathcal{N}} \mathcal{I}_N \subset \mathcal{I}$. But, since $M \in \mathcal{N}$ and $\mathcal{I}_M = \mathcal{I}$, we get

$$\bigcup_{N \in \mathcal{N}} \mathcal{I}_N = \mathcal{I}.$$

It can be shown that

$$\dim(\mathcal{I}) = \frac{1}{2}(n - k)(n - k + 1).$$

In a special case when $N = M_\Sigma$ we get invariant quadratic forms based on the weighted least squares residuals

$$\hat{\varepsilon}_\Sigma = X - P_\Sigma X = (I_n - P_\Sigma)X = M_\Sigma X,$$

where Σ is a p.d. matrix. We can write for any $A \in S^n$

$$\hat{\varepsilon}'_\Sigma A \hat{\varepsilon}'_\Sigma = X'M'_\Sigma AM_\Sigma X = X'MM'_\Sigma AM_\Sigma MX = \hat{\varepsilon}'M'_\Sigma AM_\Sigma \hat{\varepsilon},$$

since $M'_\Sigma AM_\Sigma \in \mathcal{I}$, and we see that any such quadratic form can be considered as an invariant quadratic form based on the ordinary least squares residuals and defined by the matrix $M'_\Sigma AM_\Sigma$ depending on Σ.

In general any invariant quadratic form $Q(X) = X'AX; A \in \mathcal{I}$, can be written as

$$X'AX = X'MAMX = tr(AMX(MX)') = (A, S(X)),$$

where, as before, $S(X) = (X - F\hat{\beta})(X - F\hat{\beta})' = \hat{\varepsilon}\hat{\varepsilon}'$.

Using the results already derived, we can write

$$E_\Sigma[X'AX] = tr(MAM\Sigma) = (MAM, \Sigma)$$

and, if $X \sim N_n(F\beta, \Sigma)$,

$$Cov_\Sigma(X'AX; X'BX) = 2tr(MAM\Sigma MBM\Sigma)$$

for any $A, B \in \mathcal{I}$ and any n.d. symmetric matrix Σ.

From the expression for a mean value of an invariant quadratic form it can be seen that $\tilde{g}(X) = X'AX$ is, for any $A \in \mathcal{I}$, an unbiased estimator of a linear parametric function $g(\Sigma) = tr(MAM\Sigma); \Sigma \geq 0$, of the variance-covariance components of X.

It follows from the definition of the DOOLSE and the DOWELSE, and from the result derived, that in LRMs all these estimators are given by invariant quadratic forms and thus they have desirable property that their values do not depend on the regression parameters β. This is not true for the DOOLSE and DOWELSE in an NRM, although these estimators are also defined by using the matrices S_Σ depending on residuals.

The following theorem solves the problem of finding the *locally best, invariant quadratic estimator* g_0^*, minimizing the MSE by the given covariance matrix $\Sigma = \Sigma_0$, in a class of all invariant quadratic estimators, of a parametric function g defined on the set of all p.d. covariance matrices Σ.

Theorem 1.5.1. *Let* $X \sim N(F\beta, \Sigma)$; Σ *a p.d. matrix. Let* $g(\Sigma); \Sigma \geq 0$, *be any parametric function. Then the estimator* g_0^*, *defined by*

$$g_0^*(X) = \frac{g(\Sigma_0)}{n-k+2}(X - F\hat{\beta}_{\Sigma_0})'\Sigma_0^{-1}(X - F\hat{\beta}_{\Sigma_0}),$$

is the locally, at the given p.d. covariance matrix Σ_0, *best invariant quadratic estimator of* g. *It has the MSE*

$$MSE_{\Sigma_0}[g_0^*(X)] = \frac{2g^2(\Sigma_0)}{n-k+2}.$$

Proof. See Seely (1971), Štulajter (1989).

It should be noted that the estimator g_0^* is not unbiased, since

$$E_\Sigma[g_0^*(X)] = \frac{g(\Sigma_0)}{n-k+2}tr(M'_{\Sigma_0}\Sigma_0^{-1}M_{\Sigma_0}\Sigma) = \frac{g(\Sigma_0)}{n-k+2}tr(M'_{\Sigma_0}\Sigma_0^{-1}\Sigma)$$

and for $\Sigma = \Sigma_0$ we get

$$E_{\Sigma_0}[g_0^*(X)] = \frac{g(\Sigma_0)}{n-k+2}tr(M'_{\Sigma_0}) = \frac{n-k}{n-k+2}g(\Sigma_0).$$

Example 1.5.3. Let X follow an LRM with $\Sigma_\sigma = \sigma^2 C; \sigma^2 \in (0, \infty)$, where C is a given p.d. matrix. Then

$$g_0^*(X) = \frac{\sigma_0^2}{n-k+2}(X - F\hat{\beta}_\Sigma)'\frac{C^{-1}}{\sigma_0^2}(X - F\hat{\beta}_\Sigma),$$

does not depend on σ_0^2 and thus the estimator

$$\sigma^{2*}(X) = \frac{1}{n-k+2}\hat{\varepsilon}_\Sigma C^{-1}\hat{\varepsilon}_\Sigma$$

is the *uniformly best, invariant quadratic estimator* of $g(\sigma^2) = \sigma^2; \sigma^2 \in (0, \infty)$.

Setting $C = I$ we get the equality

$$\sigma^{2*}(X) = \frac{1}{n-k+2} \sum_{i=1}^{n} (X_i - (F\hat{\beta})_i)^2 = \frac{n}{n-k+2} \hat{\sigma}^2(X),$$

where $\hat{\sigma}^2$ is the DOWELSE, or the DOOLSE, of σ^2 derived in Example 1.4.4. For σ^{2*} we have

$$MSE_{\sigma^2}[\sigma^{2*}(X)] = \frac{2\sigma^4}{n-k+2}; \sigma^2 \in (0, \infty),$$

and for the DOOLSE $\hat{\sigma}^2$:

$$MSE_{\sigma^2}[\hat{\sigma}^2(X)] = \frac{2n - 2k + k^2}{n^2} \sigma^4.$$

Since the DOOLSE $\hat{\sigma}^2$ is also an invariant estimator, its MSE must be greater than the MSE of σ^{2*}. This can easily be verified by a direct computation. Both the DOOLSE and the σ^{2*} are consistent estimators.

Example 1.5.4. Let us consider the mixed LRM considered in Example 1.4.2 with two variance components $\nu = (\sigma_1^2, \sigma_2^2)' \in (0, \infty) \times (0, \infty)$ and with matrices

$$V_1 = \begin{pmatrix} I_{n_1} & 0 \\ 0 & 0 \end{pmatrix} \text{ and } V_2 = \begin{pmatrix} 0 & 0 \\ 0 & I_{n_2} \end{pmatrix}; n_1 + n_2 = n.$$

Then the locally best, invariant estimators of variance components are given by

$$\sigma_{j,0}^{2*}(X) = \frac{\sigma_{j,0}^2}{n-k+2} \hat{\varepsilon}_{\Sigma_0}' \left(\frac{1}{\sigma_{1,0}^2} V_1 + \frac{1}{\sigma_{2,0}^2} V_2 \right) \hat{\varepsilon}_{\Sigma_0}; j = 1, 2,$$

or

$$\sigma_{1,0}^{2*}(X) = \frac{1}{n-k+2} \left[\sum_{i=1}^{n_1} (X_i - (F\beta_{\nu_0}^*)_i)^2 + \frac{\sigma_{1,0}^2}{\sigma_{2,0}^2} \sum_{i=n_1+1}^{n} (X_i - (F\beta_{\nu_0}^*)_i)^2 \right]$$

and

$$\sigma_{2,0}^{2*}(X) = \frac{1}{n-k+2} \left[\frac{\sigma_{2,0}^2}{\sigma_{1,0}^2} \sum_{i=1}^{n_1} (X_i - (F\beta_{\nu_0}^*)_i)^2 + \sum_{i=n_1+1}^{n} (X_i - (F\beta_{\nu_0}^*)_i)^2 \right].$$

Since $\beta_{\nu_0}^*$ depends only on the ratio $\sigma_{1,0}^2/\sigma_{2,0}^2$, we can also see that $\nu_0^*(X)$ depends only on this ratio. Next we have that

$$MSE_{\nu_0}[\sigma_{j,0}^{2*}(X)] = \frac{2\sigma_{j,0}^4}{n-k+2}; j = 1, 2,$$

and thus the locally best invariant quadratic estimators for $\sigma_j^2; j = 1, 2$, are consistent.

Example 1.5.5. Let $X \sim N(F\beta; \Sigma_\nu); \beta \in E^k, \nu \in \Upsilon$, where $\nu = (\sigma^2, \rho)' \in \Upsilon = (0, \infty) \times (-1, 1)$ and Σ_ν are defined as in Example 1.3.5. Then we can write $\Sigma_\nu^{-1} = \sigma^{-2} D(\rho)$ and from this we get that the locally best, invariant quadratic estimator σ_0^{2*} for a function $g(\nu) = \sigma^2; \nu \in \Upsilon$, depends only on ρ_0 and does not depend on σ_0^2. This is given by

$$\sigma_0^{2*}(X) = \frac{1}{n - k + 2} \left[(X_1 - (F\beta_0^*)_1)^2 + (X_n - (F\beta_0^*)_n)^2 \right]$$
$$+ \left[(1 + \rho_0^2) / (n - k + 2) \right] \sum_{i=2}^{n-1} (X_i - (F\beta_0^*)_i)^2$$
$$- \left[2\rho_0 / (n - k + 2) \right] \sum_{i=1}^{n-1} (X_i - (F\beta_0^*)_i) (X_{i+1} - (F\beta_0^*)_{i+1}),$$

where $\beta_0^* = \hat{\beta}_{\Sigma_{\nu_0}}$ derived in Example 1.3.5 depends only on ρ_0, and σ_0^{2*} has the MSE

$$MSE_{\nu_0}[\sigma_0^{2*}(X)] = \frac{2\sigma_0^4}{n - k + 2}.$$

For the function $g(\nu) = \rho; \nu \in \Upsilon$, we get that

$$\rho_0^*(X) = \frac{\rho_0}{\sigma_0^2} \sigma_0^{2*}(X)$$

and

$$MSE_{\nu_0}[\rho_0^*(X)] = \frac{2\rho_0^2}{n - k + 2}.$$

This estimator depends on ν_0. Both estimators σ_0^{2*} and ρ_0^* are consistent.

1.6   Unbiased Invariant Estimators

Let us consider a mixed LRM

$$X = F\beta + \varepsilon; \beta \in E^k, E[\varepsilon] = 0; E[\varepsilon\varepsilon'] = \Sigma_\nu = \sum_{j=1}^{l} \nu_j V_j; \nu \in \Upsilon.$$

Let $g(\nu) = g'\nu; \nu \in \Upsilon$, where $g \in E^l$, be a *linear parametric function* of the variance-covariance components which should be estimated on the basis of X. Let us now show how the problem of the existence of an *unbiased invariant quadratic estimator* $\tilde{g}$ of g is solved.

Let $M = I_n - P = I_n - X(X'X)^{-1}X'$ be an orthogonal projection matrix on the subspace $L(F)^\perp$ and let H be a Gramm matrix of the set $L(\mathcal{V}_M)$, with the inner product $(.,.)$, where $\mathcal{V}_M = \{MV_jM; j = 1, 2, ..., l\}$, this means

$$H_{ij} = (MV_iM, MV_jM) = tr(MV_iMV_j); i, j = 1, 2, ..., l.$$

It is clear that $L(\mathcal{V}_M)$ is a subspace of $\mathcal{I}$ and that $dim(L(\mathcal{V}_M)) = r(H) \leq l$, where the equality holds if and only if H is nonsingular. Let

$$S(X) = (X - F\hat{\beta})(X - F\hat{\beta})' = \hat{\varepsilon}\hat{\varepsilon}'.$$

The following theorem describes conditions under which a linear function g of variance-covariance components is *estimable*, which means that it has an *unbiased* invariant quadratic estimator.

Theorem 1.6.1. *A linear parametric function* $g(\nu) = g'\nu; \nu \in \Upsilon$, *where* $g \in E^l$ *is estimable iff* $g \in L(H)$. *If* g *is estimable, then the estimator*

$$\tilde{g}(X) = \sum_{k=1}^{l}(H^-g)_k(S(X), V_k) = X'M\sum_{k=1}^{l}(H^-g)_kV_kMX,$$

where H^- *is any g-inverse of* H, *is an unbiased invariant quadratic estimator of* g. *If* X *has normal distribution, then*

$$D_\nu[\tilde{g}(X)] = 2tr((M\sum_{k=1}^{l}(H^-g)_kV_kM\Sigma_\nu)^2); \nu \in \Upsilon.$$

Proof. See Štulajter (1989).

Remarks. It follows from this theorem that in the case when H is nonsingular, that means when there exists only one g-inverse H^{-1} of H, every linear parametric function g is estimable.

The matrix $M\sum_{k=1}^{l}(H^-g)_kV_kM$, which defines the unbiased estimator $\tilde{g}$ of g, belongs to the subspace $L(\mathcal{V}_M)$ of $\mathcal{I}$.

For the unbiased estimator $\tilde{g}$ it is easy to derive the following expressions:

$$\begin{aligned}
\tilde{g}(X) &= \sum_{j=1}^{l}g_j\sum_{k=1}^{l}H_{kj}^-(S(X), V_k) = \sum_{j=1}^{l}g_jX'M\sum_{k=1}^{l}H_{kj}^-V_kMX \\
&= X'M\sum_{j=1}^{l}g_j\sum_{k=1}^{l}H_{kj}^-V_kMX.
\end{aligned}$$

Let us use a symmetric Moore-Penrose g-inverse H^+ instead of, possibly monosymmetric H^-, and define the invariant quadratic forms

$$\tilde{\nu}_j(X) = \sum_{k=1}^{l} H_{jk}^+(S(X), V_k) = \sum_{k=1}^{l} H_{jk}^+ X' M V_k M X; j = 1, 2, ..., l.$$

Then one unbiased invariant quadratic estimator $\tilde{g}$ of g can be written as

$$\tilde{g}(X) = \sum_{j=1}^{l} g_j \tilde{\nu}_j(X) = g' \tilde{\nu}(X), \text{ where } \tilde{\nu}(X) = (\tilde{\nu}_1(X), ..., \tilde{\nu}_l(X))'.$$

If the parametric functions $g_j(\nu) = \nu_j; \nu \in \Upsilon$, are estimable, for example, if H is nonsingular and $H^+ = H^{-1}$, then $\tilde{\nu}_j$ are *unbiased estimators of ν_j* for $j = 1, 2, ..., l$.

It should be remarked that the expressions for $\tilde{\nu}_j$ are the same as those for the DOOLSE $\hat{\nu}_j$ for all $j = 1, 2, ..., l$, with the only exception that the matrix G^-, where $G_{ij} = (V_i, V_j)$, be replaced by the matrix H^+, where the components of H are $H_{ij} = (MV_iM, MV_jM); i, j = 1, 2, ..., l$.

A space of all unbiased invariant quadratic estimators of g can be described as follows. Since $L(\mathcal{V}_M)$ is a subspace of $\mathcal{I}$, we can write

$$\mathcal{I} = L(\mathcal{V}_M) \oplus L(\mathcal{V}_M)^\perp, \dim(\mathcal{I}) = \dim L(\mathcal{V}_M) + \dim L(\mathcal{V}_M)^\perp,$$

and every $A \in \mathcal{I}$ can be written as $A = A_V + A_V^\perp$, where A_V is a projection of A on $L(\mathcal{V}_M)$ and $A_V^\perp = A - A_V \in \mathcal{I}$ is orthogonal to $L(\mathcal{V}_M)$. It is clear that $\dim(L(\mathcal{V}_M)) = r(H) \leq l$, where the equality holds if and only if H is nonsingular, and

$$\dim(L(\mathcal{V}_M)^\perp) = \frac{1}{2}(n - k)(n - k + 1) - r(H).$$

For any $B \in L(\mathcal{V}_M)^\perp$ we get

$$
\begin{aligned}
E_\nu[X'BX] &= E_\nu[X'MBMX] = tr(MBM\Sigma_\nu) = tr(MBMM\Sigma_\nu M) \\
&= \sum_{j=1}^{l} \nu_j tr(MBMV_j) = \sum_{j=1}^{l} \nu_j (MBM, MV_jM) = 0
\end{aligned}
$$

for every $\nu \in \Upsilon$.

Using this result we get that for a fixed $A \in L(\mathcal{V}_M)$ the invariant quadratic forms $X'(A + B)X; B \in L(\mathcal{V}_M)^\perp$, are unbiased invariant quadratic estimators of the parametric function $g(\nu) = \sum_{j=1}^{l} \nu_j tr(MAMV_j); \nu \in \Upsilon$.

Thus all unbiased invariant quadratic estimators of the parametric function $g(\nu) = g'\nu; \nu \in \Upsilon$, where $g \in L(H)$, are given by

$$\tilde{g}(X) = X'M \left(\sum_{j=1}^{l} g_j \sum_{k=1}^{l} H_{jk}^+ V_k + B \right) MX; B \in L(\mathcal{V}_M)^{\perp}.$$

An estimable function g has a unique unbiased, invariant quadratic estimator g^* iff $\mathcal{I} = L(\mathcal{V}_M)$. Since g^* is unique in this case, it must be the *uniformly best, unbiased invariant quadratic estimator* of g.

For $\mathcal{I} \neq L(\mathcal{V}_M)$, the *locally best*, at the given ν_0, *unbiased invariant quadratic estimator* g_{U0}^*, minimizing the MSE at ν_0, will be described. To give this description, the following ideas are needed.

Let $\{MW_j M; j = 1, 2, ..., p\}$, where $p = \frac{1}{2}(n-k)(n-k+1) - r(H)$, be a base for $L(\mathcal{V}_M)^{\perp}$, and let us use the notation $\Sigma_0 = \Sigma_{\nu_0}$ for a given $\nu_0 \in \Upsilon$.

Let C_ν be the covariance matrix of the random vector Q with components $Q_j = X'MV_j MX; j = 1, 2, ..., l$ computed by ν and let $C_0 = C_{\nu_0}$. Let U be a random vector with components $U_j(X) = X'MW_j MX; j = 1, 2, ..., p$ and let D_0 be the covariance matrix of U computed by ν_0. Thus, if $X \sim N_n(F\beta, \Sigma_\nu)$, then

$$\begin{aligned} C_{\nu,ij} &= 2tr(MV_i M\Sigma_\nu MV_j M\Sigma_\nu); i, j = 1, 2, ..., l; \nu \in \Upsilon, \\ C_{0,ij} &= 2tr(MV_i M\Sigma_0, MV_j M\Sigma_0); i, j = 1, 2, ..., l, \end{aligned}$$

and

$$D_{0,ij} = 2tr(MW_i M\Sigma_0 MW_j M\Sigma_0); i, j = 1, 2, ..., p.$$

Let r_0 be an $p \times 1$ vector with elements

$$\begin{aligned} r_{0,j} &= Cov_{\nu_0}(X'A_g X; X'MW_j MX) \\ &= 2tr(A_g \Sigma_0 MW_j M\Sigma_0); j = 1, 2, ..., p, \end{aligned}$$

where

$$A_g = M \sum_{k=1}^{l} (H^- g)_k V_k M.$$

The estimator g_{U0}^* is given by

$$g_{U0}^*(X) = X'MA_g MX + a^{*'}U,$$

where the $p \times 1$ vector a^* is a solution of the minimizing problem

$$
\begin{aligned}
a^* &= \arg\min_{a} D_{\nu_0}[X'MA_gMX + a'U] \\
&= \arg\min_{a}\{D_{\nu_0}[X'MA_gMX] + 2a'r_0 + r_0'D_0 r_0\}.
\end{aligned}
$$

It is easy to show that

$$
a^* = -D_0^- r_0 \text{ and } D_{\nu_0}[g_{U0}^*(X)] = D_{\nu_0}[X'MA_gMX] - r_0'D_0^+ r_0.
$$

These results are formulated in the following theorem.

Theorem 1.6.2. *Let $X \sim N_n(F\beta, \Sigma_\nu)$ follow an MLRM, let $g(\nu) = g'\nu$ be an estimable parametric function defined on Υ, and let $\nu_0 \in \Upsilon$. Then the locally, at ν_0, best unbiased invariant quadratic estimator g_{U0}^* of g is given by*

$$
\begin{aligned}
g_{U0}^*(X) &= (\sum_{k=1}^{l}(H^- g)_k V_k - \sum_{k=1}^{p}(D_0^- r_0)_k W_k, S(X)) \\
&= X'M\sum_{k=1}^{l}(H^- g)_k V_k MX - X'M\sum_{k=1}^{p}(D_0^- r_0)_k W_k MX
\end{aligned}
$$

and has the MSE

$$
MSE_{\nu_0}[g_{U0}^*(X)] = g'H^+ C_0 H^+ g - r_0'D_0^+ r_0.
$$

Remark. It can be easily verified that the estimator $\tilde{g}(X)$ given in Theorem 1.6.1. can be written as

$$
\tilde{g}(X) = (A_g, S(X))
$$

and

$$
\begin{aligned}
MSE_\nu[\tilde{g}(X)] &= D_\nu[g'\tilde{\nu}(X)] \\
&= g'H^+ C_\nu H^+ g; \nu \in \Upsilon.
\end{aligned}
$$

From this expression we have that the covariance matrix $Cov_\nu(\tilde{\nu}(X))$ of the estimator $\tilde{\nu}(X)$ is given by

$$
Cov_\nu(\tilde{\nu}(X)) = H^+ C_\nu H^+; \nu \in \Upsilon.
$$

Example 1.6.1. Let us consider a classical LRM with covariance matrices $\Sigma_\sigma = \sigma^2 I_n; \sigma^2 \in (0, \infty)$. Then the Gramm matrix H is 1×1 and is equal to $tr((MI_nM)^2) = tr(M)$ and thus $H^{-1} = (n-k)^{-1}$. For all matrices $MW_jM; j = 1, 2, ..., p$, belonging to $L(\mathcal{V}_M)^\perp$ and for any positive σ_0^2, we have

$$
\begin{aligned}
d_{0,j} &= 2\sigma_0^4 tr(A_g MW_jM) = 2\sigma_0^4 \sum_{k=1}^{l}(H^- g)_k tr(V_k MW_jM) \\
&= 2\sigma_0^4 \sum_{k=1}^{l}(H^- g)_k(MV_kM, MW_jM) = 0; j = 1, 2, ..., p,
\end{aligned}
$$

and thus the locally best, unbiased invariant quadratic estimator σ_{U0}^{2*} does not depend on σ_0 and is *the uniformly best, unbiased invariant estimator of σ^2* given by

$$
\sigma_U^{2*}(X) = \frac{1}{n-k}\sum_{i=1}^{n}((X_i - F\hat\beta)_i)^2.
$$

It has the variance

$$
D_\sigma[\sigma_U^{2*}(X)] = \frac{2\sigma^4}{n-k} \text{ for every } \sigma^2 \in (0, \infty).
$$

Thus for σ^2 we have three invariant estimators: the MLE $\hat\sigma^2$, the uniformly best, invariant quadratic estimator σ^{2*}, and the uniformly best, unbiased invariant quadratic estimator σ_U^{2*}. All these estimators are only different multiples of the statistics

$$
(I_n, S(X)) = tr(S(X)) = \sum_{i=1}^{n}((X_i - (F\hat\beta)_i)^2.
$$

σ^{2*} and $\hat\sigma^2$ are biased, σ^{2*} has minimal MSE. We can compare the MSEs of $\hat\sigma^2$ and σ_u^{2*}. It is easy to show that

$$
MSE_{\sigma^2}[\hat\sigma^2(X)] = \frac{2n - 2k + k^2}{n^2}\sigma^4 \le MSE_{\sigma^2}[\sigma_U^{2*}(X)] = \frac{2}{n-k}\sigma^4
$$

for all positive σ^2 and all $n \ge k + 1$.

Thus, from the point of view of MSEs the uniformly best unbiased invariant quadratic estimator σ_U^{2*} of σ^2 is not admissible. But the difference between the MSEs of these three estimators is very small and, for mainly big values of n, these estimators can be regarded as equivalent. For other reasons the unbiased estimator σ_U^{2*} is often used in statistical inference.

Example 1.6.2. Let $X \sim N_3(F\beta, \Sigma_\nu); \nu \in \Upsilon$, where $F = j_3$, and let $\Sigma_\nu = \sum_{j=1}^{3} \sigma_j^2 V_j; \sigma_j^2 \geq 0$, where

$$V_1 = \begin{pmatrix} 1 & 0 & 0 \\ 0 & 0 & 0 \\ 0 & 0 & 0 \end{pmatrix}, V_2 = \begin{pmatrix} 0 & 0 & 0 \\ 0 & 1 & 0 \\ 0 & 0 & 0 \end{pmatrix}, V_3 = \begin{pmatrix} 0 & 0 & 0 \\ 0 & 0 & 0 \\ 0 & 0 & 1 \end{pmatrix}.$$

The matrices $V_j; j = 1, 2, 3$, are orthonormal. Next we have

$$M = I_3 - P = \frac{1}{3} \begin{pmatrix} 2 & -1 & -1 \\ -1 & 2 & -1 \\ -1 & -1 & 2 \end{pmatrix}$$

and the Gramm matrix H of matrices $MV_jM; j = 1, 2, 3$, is given by

$$H_{ij} = tr(MV_iMV_j) = M_{ij}^2; i, j = 1, 2, 3,$$

or

$$H = \frac{1}{9} \begin{pmatrix} 4 & 1 & 1 \\ 1 & 4 & 1 \\ 1 & 1 & 4 \end{pmatrix}.$$

We see that H is nonsingular and thus every linear parametric function g is estimable. Next we have

$$\dim(L(\mathcal{V}_M)^\perp) = \frac{1}{2}(n-k)(n-k+1) - r(H) = 0$$

and thus every estimable function $g(\nu) = g'\nu; \nu \in \Upsilon$, has only one, and thus the uniformly best, unbiased invariant quadratic estimator g_U^*.

This estimator is given by

$$g_U^*(X) = g'\nu_U^*(X),$$

where

$$\nu_U^*(X) = (\sigma_{U,1}^{2*}(X), \sigma_{U,2}^{2*}(X), \sigma_{U,3}^{2*}(X))'$$

and

$$\sigma_{U,j}^{2*}(X) = \prod_{i \neq j}(X_j - X_i).$$

Every component of $\sigma_{U,j}^{2*}; j = 1, 2, 3$, can be written as an invariant quadratic form, for example,

$$\sigma_{U,1}^{2*}(X) = X' \frac{1}{2} \begin{pmatrix} 2 & -1 & -1 \\ -1 & 0 & 1 \\ -1 & 1 & 0 \end{pmatrix} X,$$

and using this and similar expressions we can easily compute that

$$D_\nu[\sigma^{2*}_{U,j}(X)] = 2\sigma^4_j + \sigma^2_1\sigma^2_2 + \sigma^2_1\sigma^2_3 + \sigma^2_2\sigma^2_3; j = 1,2,3.$$

It should be remarked that the mixed LRM considered in this example is overparametrized, since the number of parameters, $k+l = 4$, of the model is greater than the number of observations, $n = 3$. This could be a reason for the fact that the estimators $\sigma^{2*}_{U,j}; j = 1,2,3$, can take negative values. This undesirable property can also be caused by the unbiasedness requirements.

Let us also consider the DOOLSE $\hat{\nu}$. The DOOLSEs $\hat{\sigma}^2_j$ for σ^2_j are given by

$$\hat{\sigma}^2_j(X) = (X_j - \bar{X})^2; j = 1,2,3.$$

These are invariant quadratic forms, for example,

$$\hat{\sigma}^2_1(X) = X'\frac{1}{9}\begin{pmatrix} 4 & -2 & -2 \\ -2 & 1 & 1 \\ -2 & 1 & 1 \end{pmatrix} X,$$

and we can easily compute that

$$E_\nu[\hat{\sigma}^2_1] = \frac{1}{9}(4\sigma^2_1 + \sigma^2_2 + \sigma^2_3)$$

and

$$D_\nu[\hat{\sigma}^2_1(X)] = \frac{2}{81}[16\sigma^4_1 + \sigma^4_2 + \sigma^4_3 + 8(\sigma^2_1\sigma^2_2 + \sigma^2_1\sigma^2_3) + \sigma^2_2\sigma^2_3].$$

It is easy to show, using similar equalities for other components of the DOOLSE $\hat{\nu}$, that we get, for $j = 1,2,3$,

$$MSE_\nu[\hat{\sigma}^2_j(X)] = \frac{1}{81}\left[57\sigma^4_j + 3\sum_{i\neq j}\sigma^4_i + 6(\sigma^2_1\sigma^2_2 + \sigma^2_1\sigma^2_3 + \sigma^2_2\sigma^2_3)\right].$$

These MSEs can be compared with those of the uniformly best, unbiased invariant quadratic estimators σ^{2*}_j which are

$$MSE_\nu[\sigma^{2*}_j(X)] = 2\sigma^4_j + \sigma^2_1\sigma^2_2 + \sigma^2_1\sigma^2_3 + \sigma^2_2\sigma^2_3; j = 1,2,3.$$

It can be seen that the MSEs of the DOOLSEs $\hat{\sigma}^2_j$ are, on a relatively large subset of the parametric space Υ, smaller then the MSEs of the uniformly best, unbiased invariant quadratic estimators σ^{2*}_j for all $j = 1,2,3$.

Example 1.6.3. Let $X \sim N_n(j_n\beta; \Sigma_\nu); \nu \in \Upsilon$, where $\beta \in E^1$, and the covariance matrices Σ_ν are given by

$$\Sigma_\nu = \sum_{j=1}^n \sigma_j^2 V_j; \nu = (\sigma_1^2, ..., \sigma_n^2)' \in \Upsilon = (0, \infty)^n,$$

where $V_{j,ik} = 1$ for $i = k = j$ and 0 elsewhere; $j = 1, 2, ..., n$. Thus components X_i of X are independent random variables with $E_\beta[X_i] = \beta; \beta \in E^1$, and $D[X_i] = \sigma_i^2; i = = 1, 2, ..., n$. This mixed LRM is called *heteroscedastic*. In Example 1.6.2 we assumed that $n = 3$.

In the general case

$$H_{ij} = tr(MV_iMV_j) = M_{ij}^2 = \left(I_n - \frac{1}{n}J_n\right)_{ij}^2; i, j = 1, 2, ..., n,$$

and we get

$$H = \frac{1}{n^2}\begin{pmatrix} (n-1)^2 & 1 & ... & 1 \\ 1 & (n-1)^2 & ... & 1 \\ . & . & ... & . \\ 1 & 1 & ... & (n-1)^2 \end{pmatrix}, r(H) = n,$$

and thus every σ_j^2 has an unbiased estimator $\tilde{\sigma}_j^2; j = 1, 2, ..., n$. Next we have

$$\dim(L(\mathcal{V}_M)^\perp) = \frac{1}{2}(n-k)(n-k+1) - r(H) = \frac{1}{2}(n^2 - 3n)$$

which is a positive number, if $n \geq 4$. We conclude that for $n \geq 4$ there is no longer the uniformly best, unbiased invariant quadratic estimator ν_U^* of ν. An unbiased invariant quadratic estimator can be found using Theorem 1.6.1.

The matrix H can be written in the form

$$H = \frac{1}{n^2}[((n-1)^2 - 1)I_n + J_n]$$

and, using the equality,

$$(aI_n + bJ_n)^{-1} = \frac{1}{a}\left(I_n - \frac{b}{a+nb}J_n\right),$$

which is known from matrix algebra, we can write

$$H^{-1} = \frac{n^2}{n^2 - 2n}\left(I_n - \frac{1}{n^2 - n}J_n\right) = \frac{n^2}{(n^2 - 2n)(n^2 - n)}((n^2 - n)I_n - J_n).$$

Using this expression we can derive unbiased estimators $\tilde{\sigma}_j^2$ of σ_j^2 for every $j = 1, 2, ..., n$.

These are given by

$$\tilde{\sigma}_j^2(X) = \frac{n^2(n^2 - n)}{(n^2 - 2n)(n^2 - n)}(X_j - \bar{X})^2 - \frac{n^2}{(n^2 - 2n)(n^2 - n)}\sum_{i=1}^{n}(X_i - \bar{X})^2.$$

For large values of n the unbiased estimators $\tilde{\sigma}_j^2$ are approximately equal to the DOOLSEs $\hat{\sigma}_j^2$. It can be easily shown that the DOOLSE $\hat{\nu}$ for ν has components $\hat{\sigma}_j^2$ which are given by

$$\hat{\sigma}_j^2(X) = (X_j - \bar{X})^2; j = 1, 2, ..., n.$$

The estimator $\hat{\sigma}_1^2(X)$ can be written as an invariant quadratic form $X'A_1X$, where the matrix A_1 is given by

$$A_1 = \frac{1}{n^2}\begin{pmatrix} (n-1)^2 & -(n-1) & \cdots & -(n-1) \\ -(n-1) & & & \\ \cdot & & J_{n-1} & \\ -(n-1) & & & \end{pmatrix}.$$

Using this, and similar expressions for $\hat{\sigma}_j^2$, we can easily get that

$$E_\nu[\hat{\sigma}_j^2(X)] = \left(\frac{n-1}{n}\right)^2 \sigma_j^2 + \frac{1}{n^2}\sum_{i\neq j}\sigma_i^2; j = 1, 2, ..., n,$$

and, using the fact that $tr(A_1V_iA_1V_j) = A_{1,ij}^2; i, j = 1, 2, ..., n$, we get

$$D_\nu[\hat{\sigma}_j^2(X)] = 2\left[\left(\frac{n-1}{n}\right)^4 \sigma_j^4 + \frac{1}{n^4}\sum_{i\neq j}\sigma_i^4 + \left(\frac{n-1}{n^2}\right)^2 \sum_{i\neq k}\sigma_i^2\sigma_k^2\right]; \nu \in \Upsilon.$$

We see from these expressions that the DOOLSEs $\hat{\sigma}_j^2; j = 1, 2, ..., n$, are not unbiased, they are only asymptotically unbiased. Their variances do not converge to zero as n tends to infinity and thus they are not consistent estimators. The asymptotic MSEs of $\hat{\sigma}_j^2$ are equal to the asymptotic MSEs of $\tilde{\sigma}_j^2$ and are equal to $2\sigma_j^4; j = 1, 2, ..., n.$

Example1.6.4. Let us consider the mixed LRM studied in Example 1.5.4 with covariance matrices

$$\Sigma_\nu = \sigma_1^2 V_1 + \sigma_2^2 V_2; \sigma_1^2, \sigma_2^2 \in (0, \infty),$$

where

$$V_1 = \begin{pmatrix} I_{n_1} & 0 \\ 0 & 0 \end{pmatrix} \text{ and } V_2 = \begin{pmatrix} 0 & 0 \\ 0 & I_{n_2} \end{pmatrix}; n_1 + n_2 = n.$$

Let us write the projection matrix $M = I_n - P$ in the block form

$$M = \begin{pmatrix} M_1 & M_{1,2} \\ M_{1,2}' & M_2 \end{pmatrix},$$

where M_1 is $n_1 \times n_1$ and M_2 is $n_2 \times n_2$. Then we get

$$MV_1 = \begin{pmatrix} M_1 & 0 \\ M_{1,2}' & 0 \end{pmatrix}, MV_2 = \begin{pmatrix} 0 & M_{1,2} \\ 0 & M_2 \end{pmatrix},$$

$$MV_1MV_1 = \begin{pmatrix} M_1^2 & 0 \\ M_{1,2}'M_1 & 0 \end{pmatrix}, MV_1MV_2 = \begin{pmatrix} 0 & M_1M_{1,2} \\ 0 & M_{1,2}'M_{1,2} \end{pmatrix}$$

and

$$MV_2MV_2 = \begin{pmatrix} 0 & M_{1,2}M_2 \\ 0 & M_2^2 \end{pmatrix}.$$

Using these equalities we can write

$$H = \begin{pmatrix} tr(M_1^2) & tr(M_{1,2}'M_{1,2}) \\ tr(M_{1,2}'M_{1,2}) & tr(M_2^2) \end{pmatrix}.$$

In a special case, when $F = j_n$, we get

$$\begin{aligned} H &= \begin{pmatrix} n_1^2\left((n-1)/n\right)^2 & n_1n_2/n^2 \\ n_1n_2/n^2 & n_2^2\left((n-1)/n\right)^2 \end{pmatrix} \\ &= \frac{1}{n^2}\begin{pmatrix} n_1^2(n-1)^2 & n_1n_2 \\ n_1n_2 & n_2^2(n-1)^2 \end{pmatrix} \end{aligned}$$

and

$$\begin{aligned} H^{-1} &= \frac{n^2}{n_1n_2[n_1n_2(n-1)^4 - 1]}\begin{pmatrix} n_2^2(n-1)^2 & -n_1n_2 \\ -n_1n_2 & n_1^2(n-1)^2 \end{pmatrix} \\ &= \frac{1}{n_1n_2(n-1)(n-2)}\begin{pmatrix} n_2(n_2+n(n-2)) & -n_1n_2 \\ -n_1n_2 & n_1(n_1+n(n-2)) \end{pmatrix}. \end{aligned}$$

Thus every σ_j^2 has an unbiased invariant quadratic estimator $\tilde{\sigma}_j^2; j = 1, 2$, described in Theorem 1.6.1. These are given by

$$\begin{aligned} \tilde{\sigma}_1^2(X) &= \frac{n_2 + n(n-2)}{(n-1)(n-2)}\hat{\sigma}_1^2(X) - \frac{1}{(n-1)(n-2)}\hat{\sigma}_2^2(X), \\ \tilde{\sigma}_2^2(X) &= \frac{n_1 + n(n-2)}{(n-1)(n-2)}\hat{\sigma}_2^2(X) - \frac{1}{(n-1)(n-2)}\hat{\sigma}_1^2(X), \end{aligned}$$

where $\hat{\sigma}_j^2$ are the DOOLSEs of $\sigma_j^2; j = 1, 2$ given by

$$\hat{\sigma}_1^2(X) = \frac{1}{n_1} \sum_{i=1}^{n_1} (X_i - \bar{X})^2,$$

$$\hat{\sigma}_2^2(X) = \frac{1}{n_2} \sum_{i=n_1+1}^{n} (X_i - \bar{X})^2.$$

We can see from the expressions for unbiased estimators $\tilde{\sigma}_j^2$ that they are, for large n, approximately equal to the DOOLSEs $\hat{\sigma}_j^2; j = 1, 2$.

The DOOLSEs $\hat{\sigma}_j^2; j = 1, 2$ can be written as invariant quadratic forms, namely $\hat{\sigma}_1^2(X) = X'A_1X$, where

$$A_1 = \frac{1}{n_1} \left\{ \begin{pmatrix} I_{n_1} & 0 \\ 0 & 0 \end{pmatrix} - \frac{1}{n} \left[\begin{pmatrix} j_{n_1} \\ 0 \end{pmatrix} j_n' + j_n \begin{pmatrix} j_{n_1} \\ 0 \end{pmatrix}' \right] + \frac{n_1}{n} J_n \right\}.$$

From this expression we get, after some computation,

$$E_\nu \left[\hat{\sigma}_1^2(X) \right] = tr(A_1V_1) + \sigma_2^2 tr(A_1V_2) = \frac{n-2}{n}\sigma_1^2 + \frac{n_2}{n}\sigma_2^2; \nu \in \Upsilon,$$

and we see that the DOOLSE $\hat{\sigma}_1^2$ is an *asymptotically unbiased, invariant quadratic estimator* of σ_1^2 if $n_2/n \to 1$.

For a normally distributed vector X we get

$$D_\nu \left[\hat{\sigma}_1^2(X) \right] = 2tr(A_1\Sigma_\nu A_1\Sigma_\nu) = 2 \sum_{i=1}^{n} \sum_{j=1}^{n} \sigma_i^2 \sigma_j^2 tr(A_1V_iA_1V_j).$$

Next we have

$$A_1V_1 = \frac{1}{n_1}I_{n_1} + \frac{n_1 - 2n}{n^2 n_1} J_{n_1}, \quad A_1V_2 = \frac{1}{n^2} J_{n_2},$$

$$(A_1V_1)^2 = \frac{1}{n_1}I_{n_1} - 2\frac{2n - n_1}{n^2 n_1} J_{n_1} + \left(\frac{2n - n_1}{n^2 n_1} \right)^2 J_{n_1},$$

$$A_1V_1A_1V_2 = 0 \text{ and } (A_1V_2)^2 = \frac{n_2}{n} J_{n_1}.$$

Using these expressions it is easy to show that

$$D_\nu \left[\hat{\sigma}_1^2(X) \right] = \left[\frac{1}{n_1} - 2\frac{2n_2 + n_1}{n^2 n_1} + \left(\frac{2n_2 + n_1}{n^2} \right)^2 \right] \sigma_1^4 + \frac{n_2^2}{n^4} \sigma_2^4$$

$$= \left[\frac{1}{n_1} \left(\frac{n-2}{n} \right)^2 + \frac{2}{n^2} \frac{n_1 - 2}{n_1} + \left(\frac{2n - n_1}{n^2} \right)^2 \right] \sigma_1^4 + \frac{n_2^2}{n^4} \sigma_2^4.$$

The square of the bias is

$$\left(E_\nu \left[\hat{\sigma}_1^2(X) - \sigma_1^2 \right] \right)^2 = \frac{1}{n^2} \left(2\sigma_1^2 + \frac{n_2}{n} \sigma_2^2 \right)^2 ; \nu \in \Upsilon.$$

It can be seen from these expressions that the DOOLSE $\hat{\sigma}_1^2$ is a consistent estimator of σ_1^2. Similar equalities, with n_1 replaced by n_2 and n_2 replaced by n_1 can be derived for the DOOLSE $\hat{\sigma}_2^2$. This estimator is again a consistent estimator of σ_2^2 if $n_2/n \to 1$.

In the last example of this chapter we give an example of a mixed LRM which is very important for time series and which shows that there do not exist unbiased invariant quadratic estimators of variance-covariance components.

Example1.6.5. Let X follows the mixed LRM considered in Example 1.4.3 with $n = 3$, which means

$$X \sim N_3(j_3\beta; \Sigma_\nu); \beta \in E^3, \Sigma_\nu = \nu_0 I_3 + \sum_{j=1}^{2} \nu_j V_j; \nu \in \Upsilon,$$

where

$$V_0 = I, V_j = \begin{pmatrix} 0 & I_{3-j} \\ 0 & 0 \end{pmatrix} + \begin{pmatrix} 0 & 0 \\ I_{3-j} & 0 \end{pmatrix} ; j = 1, 2.$$

It is easy to show that

$$M = \frac{1}{3} \begin{pmatrix} 2 & -1 & -1 \\ -1 & 2 & -1 \\ -1 & -1 & 2 \end{pmatrix}$$

and that the Gramm matrix H is given by

$$H = \frac{2}{9} \begin{pmatrix} 9 & -6 & -3 \\ -6 & 8 & -2 \\ -3 & -2 & 5 \end{pmatrix}.$$

H is a singular matrix, $r(H) = 2$, and the vectors $(1, 0, 0)'$, $(0, 1, 0)'$, and $(0, 0, 1)'$ do not belong to $L(H)$. Thus for every component $\nu_j; j = 0, 1, 2$, of ν there *does not exist an unbiased invariant quadratic estimator*. The same is true for $X \sim N_n(F\beta; \Sigma_\nu); \beta \in E^n, \nu \in \Upsilon$, where

$$\Sigma_\nu = \nu_0 I_n + \sum_{j=1}^{n-1} \nu_j V_j; \nu \in \Upsilon,$$

with

$$V_j = \begin{pmatrix} 0 & I_{n-j} \\ 0 & 0 \end{pmatrix} + \begin{pmatrix} 0 & 0 \\ I_{n-j} & 0 \end{pmatrix} ; j = 1, 2, ..., n - 1.$$

There is no unbiased invariant quadratic estimator for ν. As we shall show later, the DOOLSEs

$$\hat{\nu}_j(X) = \frac{1}{n-j} \sum_{i=1}^{n-j} (X_{i+j} - (F\hat{\beta})_{i+j})(X_i - (F\hat{\beta})_i)$$

are biased estimators of ν_j, but they have, in many cases, good asymptotic properties. They are consistent estimators of ν_j for all fixed j. For this reason the DOOLSEs $\hat{\nu}_j$ are often used in the statistical inference of time series.

2

Random Processes and Time Series

Random processes and their special types, time series, are used in many fields of human life. They serve as models for real processes which are of random character, that is for processes randomly changing in time. As an example we can give the changes in temperature of air observed in some meteorological laboratory, changes level of a river, in the consumption of electrical energy in some town observed continuously during some time interval, or the heart action of a patient recorded by his ECG, some other numerical parameters of a patient observed during his stay in hospital. To this type also belongs production of some company, recorded by days or months. Another kind of random process can be represented by the measurements of the diameter of a shaft along its length, or by the measurement of consumption of a gasoline by a car at different speeds. In these last two examples we can substitute the real time by some other "time" parameter, for example, by the speed of the car.

2.1 Basic Notions

A *random process* X is defined as a set of random variables $X(t); t \in T$, where the parameter t is called *time* and T is a subset of the real line; the notation $X = \{X(t); t \in T\}$ is often used to denote a random process X. There are two main types of random processes: *random processes with continuous time*, where it is assumed that T is a union of intervals on the real line, especially $T = (-\infty, \infty) = E^1$, and *random processes with discrete time*, or *time series*, where it is assumed that the set T contains at most countably many points, especially, T is equal to the set of all integers, $Z = \{..., -2, -1, 0, 1, 2, ...\}$. The probabilistic theory of random processes was developed mainly for these last two cases when $T = E^1$ and $T = Z$.

A typical feature of the theory of random processes is that an infinite number of random variables is considered. In a special case when the set T is finite, the notion of a random process can be identified with the notion of a random vector. That is, the theory of random processes is a generalization of a theory of random vectors.

A real function $x = x(t); t \in T$, defined on T, where $x(t)$ is a realization of $X(t)$ at time t, is called *a realization of* X. A set of all realizations of X will be denoted by $\mathcal{X}$. In the probability theory of random processes it is proved that a random process X is uniquely described by its probability distribution defined on subsets of $\mathcal{X}$. But we shall not deal with these problems.

In the practical applications of the theory of random processes we assume that we observe a random process on some interval of a finite length, or on a discrete set containing a finite number of points. Let us denote this set of times, in both cases, by T_O. Then the random process $X_O = \{X(t); t \in T_O\}$ will be called an *observation of* $X = \{X(t); t \in T\}$ where we assume that $T_O \subset T$. A real function $x = x(t); t \in T_O$, is called a *realization of an observation of a random process* X. It should be remarked that the observations of random processes with continuous time are often "digitalized" and we take into account only observations at some points $t_1, ..., t_n \in T$. If the distance between the points $t_i; i = 1, 2, ..., n$, is constant, then we shall use for an observation X_O the notation $X_O = (X(1), X(2), ..., X(n))'$. The same notation will also be used for a finite observation of a random process X with discrete time if $T_O = \{t_{i+1}, t_{i+2}, ..., t_{i+n}\}$ for some i. Usually we set $i = 0$ and $t_i = i; i = 1, 2, ..., n$. A random process X is *called Gaussian, or normal*, if every finite observation X_O of X has a multivariate normal distribution.

It is clear that in real life there is available only one finite realization of an observation of a real random process. It is typical for many random processes that they cannot be repeated, since they are running in time with no replication. This is an important feature of random processes which has a big influence on the statistical inference of random processes.

The other important feature of random processes is that the random variables $X(t)$ are, for different times $t \in T$, not independent random variables. It is typical that $X(s)$ and $X(t)$ are, for different times s and t belonging to T, correlated.

All these main features of random processes should be included in statistical and probabilistic models describing the behavior of random processes, or used in the statistical inference of random processes. One important statistical problem of random processes is the problem of the prediction of future values of a random process based on its observation.

The main characteristics of a random process are the mean value and covariance function of a random process X. A real function $m(.)$ defined by

$$m(t) = E\left[X(t)\right]; t \in T,$$

is called the *mean value* of X. The mean value of a random process X is said to be a *characteristic of location* of X. This means that the realizations $x(.)$ of X lie about its mean value $m(.)$. This is an analogy to the meaning of the mean value of a random variable or of a mean value of a random vector. Theoretically, any real function defined on T can be regarded as a mean value of some random process X. But, as we shall see later, there are some reasonable classes of functions which can be considered as the mean values of random processes.

Another analogy is the notion of the covariance function of a random process to the notion of the covariance matrix of a random vector. A real function $R(.,.)$ defined on $T \times T$ by

$$R(s,t) = Cov(X(s); X(t)); s, t \in T,$$

is called the *covariance function* of a random process X. The covariance function of X, as well as the covariance matrix of some random vector, contains information about linear dependences between the coordinates of a random process X. Thus this is a base for solving the linear statistical problems of random processes. It follows directly from the definition of a covariance function that it is a symmetric function, $R(s,t) = R(t,s); s,t \in T$, and that $R(t,t) = D[X(t)]; t \in T$. Using the Schwarz inequality we get

$$|R(s,t)| \leq (R(s,s)R(t,t))^{\frac{1}{2}}; s, t \in T.$$

Every covariance function is *nonnegative definite,* this means that

$$\sum_{i=1}^{n} \sum_{j=1}^{n} a_i a_j R(t_i, t_j) \geq 0$$

for any $a = (a_1, ..., a_n)' \in E^n; t_1, ..., t_n \in T$ and for any positive integer n.

It is possible to show that every nonnegative definite symmetric function $R(.,.)$, for which the Schwarz inequality holds, is a covariance function of some random process X. Although there is a large class of possible covariance functions available, it is reasonable, in solving a statistical problem of a random process, to choose some model for its covariance function. As we shall see, this model can not only be parametric, linear, or nonlinear, but also nonparametric.

2.2 Models for Random Processes

We shall first describe some models which can be used for both random processes with continuous time and time series. The basic model for random process X is the *additive model.* It is defined by

$$X(t) = m(t) + \varepsilon(t); t \in T,$$

where it is assumed that the random process $\varepsilon = \{\varepsilon(t); t \in T\}$ has a mean value equal to zero and thus $m(.)$ is the mean value of X. Next it is assumed that ε has a covariance function $R(.,.)$ which is equal to the covariance function of X. In the additive model the covariance function $R(.,.)$ does not depend on $m(.)$.

A *multiplicative model* is defined by

$$X(t) = m(t)\varepsilon(t); t \in T,$$

where it is assumed that $E\left[\varepsilon(t)\right] = 1$ for all $t \in T$ and that ε has a covariance function $R^\varepsilon(.,.)$. Thus

$$E[X(t)] = m(t)$$

and

$$R(s,t) = Cov(X(s); X(t)) = m(s)m(t)R^\varepsilon(s,t); s,t \in T.$$

It is seen that in a multiplicative model the covariance function of X depends on the mean value, and for large values of the mean value of X there can also be large values of the variance $D[X(t)] = m^2(t)R^\varepsilon(t,t)$ of $X(t)$. This property is natural in some real random processes. In this book we shall deal mainly with additive models.

First we shall consider some commonly used models for the mean value of a random process. Every such model is given by a set $\mathcal{M}$ of possible mean values $m(.)$ of $X(.)$.

The random process X with $E[X(t)] = m(t) = 0; t \in T$, is called *centered* and a random process with a constant mean value $m(t) = \beta; t \in T, \beta \in E^1$, is called *stationary in mean value*.

For the mean $m(.)$ value of X the following models are also often used in the statistical modeling of random processes. The first, and most commonly used, is an LRM. X fulfills an LRM if

$$X(t) = m(t) + \varepsilon(t); t \in T, m(.) \in \mathcal{M},$$

where

$$\mathcal{M} = \left\{ m_\beta(t) = \sum_{i=1}^{k} \beta_i f_i(t); t \in T; \beta \in E^k \right\}$$

and where the *regression parameter* $\beta = (\beta_1, ..., \beta_k)'$ can be any vector from E^k and $f_1, ..., f_k$ are given known functions. The random process ε is called *noise* and it is assumed that

$$E[\varepsilon(t)] = 0,$$

$$Cov(X(s); X(t)) = Cov(\varepsilon(s); \varepsilon(t)) = R(s,t); s,t \in T.$$

A random process X which is stationary in an (unknown) mean value can be considered as an LRM with $k = 1$ and $f_1(t) = 1$ for all t.

As another special case of an LRM we give

$$m_\beta(t) = \beta_1 + \beta_2 t; t \in T; \beta = (\beta_1, \beta_2)' \in E^2.$$

The component β_1 of β is called an *intercept* and β_2 is called a *slope* of the mean value, also called a *linear trend*. Random processes with linear trends can be used to model economical processes, or some processes observed on a relatively small time interval, where a mean value can be approximated by a line.

A random process X is said to exhibit a *polynomial trend* if

$$m_\beta(t) = \sum_{i=1}^{k} \beta_i t^{i-1}; t \in T; \beta = (\beta_1, ..., \beta_k)' \in E^k, k \geq 3.$$

As an example of the use of such a polynomial trend is the mean value of the consumption of gasoline of a given car depending on the speed of the car. This mean value can be described by a second-order polynomial, or by a *quadratic trend*

$$m_\beta(t) = \beta_1 + \beta_2 t + \beta_3 t^2; t \in T, \beta = (\beta_1, \beta_2, \beta_3)' \in E^3.$$

Here t is the velocity of the car. It should be remarked that this random process can be repeated independently on p different cars of the same kind and make. Such repeated observations can be described by using an MLRM. This is given by

$$X_i(t) = \beta_1^i + \beta_2^i t + \beta_3^i t^2 + \varepsilon_i(t); t \in T, \beta^i = (\beta_1^i, \beta_2^i, \beta_3^i)' \in E^3; i = 1, 2, ..., p,$$

where $X_i(t)$ is a measurement on the ith car and $\varepsilon_i(t); i = 1, 2, ..., p$, are assumed to be independent noises with mean values equal to zero and with the same covariance function $R(.,.)$.

Another important LRM which can be successfully used in many practical applications has a mean value of the form

$$m_\beta(t) = \beta_1 + \beta_2 t + \sum_{i=1}^{k} (\beta_i^1 \cos \lambda_i^0 t + \beta_i^2 \sin \lambda_i^0 t); t \in T,$$

where $k \geq 1, \lambda_i^0; i = 1, 2, ..., k$, are the known *frequencies* λ of goniometric functions $\cos \lambda t, \sin \lambda t; t \in T$, and $\beta = (\beta_1, \beta_2, \beta_1^1, ..., \beta_k^1, \beta_1^2, ..., \beta_k^2)' \in E^{2k+2}$. The parameters $\beta_i^1, \beta_i^2; i = 1, 2, ..., k$, are called the *amplitudes* of the goniometric functions.

The preceding LRM, without the linear trend part, is called the *qauasiperiodic* LRM. It consists of a linear combination of the periodic goniometric functions $\cos \lambda t, \sin \lambda t; t \in T$, which are periodic with the period $T_p = 2\pi/\lambda$. A frequency λ can be any real number if T is an interval on the real line, that is, if X is a random process with continuous time. For time series with

$T = Z = \{..., -2, -1, 0, 1, 2, ...\}$, the frequencies λ can take values only from the closed interval $\langle -\pi, \pi \rangle$.

As an example of the use of a quasiperiodic LRM we can give observations of meteorological events, for example, temperature, humidity of the air, or the level of a river. When we have monthly observations, then periodicity of the length $T_p = 12$ can be expected and we have to use basic goniometric functions $\cos \lambda t, \sin \lambda t; t \in T$, with the frequency $\lambda = \lambda_1 = 2\pi/12 = \pi/6$.

There can also be other latent frequencies which should be included in the mean value of the observed random processes. In the following chapter we give a method based on a periodogram which enables us to discover such frequencies from a realization of the finite observation of a random process.

Quasiperiodic LRMs with a linear trend can be used for modeling time series describing the consumption of some nonalcoholic drink observed quarterly. It can be assumed that this random process is periodic with period $T_p = 4$ and that there is some linear trend describing a tendency of this consumption. A slope of this linear trend says weather the consumption is increasing, or decreasing, during the time of observation.

By choosing functions $f_i; i = 1, 2, ..., k$, for an LRM we should take into account the assumption that the mean value of a random process is changing slowly in time, and thus we try to take smooth functions to construct a suitable LRM. Goniometric functions with small frequencies, and thus with a big periods, fulfill this requirement. From the other side, a noise ε is expected to change rapidly in time and thus can be represented by goniometric functions with high frequencies, that with small periods.

In many applications LRMs are not satisfactory for describing a random process. In these cases we can use an NRM. This is given by

$$X(t) = m(t) + \varepsilon(t); t \in T, m(.) \in \mathcal{M},$$

where

$$\mathcal{M} = \{m_\gamma(t); t \in T, \gamma \in \Gamma\},$$

and where Γ is a *parametric space*. In NRMs the dependence of m_γ on γ is nonlinear.

A quasiperiodic LRM with unknown amlitudes and frequencies is an example of an NRM. This is given by

$$m_\gamma(t) = \sum_{i=1}^{k} (\beta_i^1 \cos \lambda_i t + \beta_i^2 \sin \lambda_i t); t \in T,$$

where $\gamma = (\beta_1^1, ..., \beta_k^1, \beta_1^2, ..., \beta_k^2, \lambda_1, ..., \lambda_k)' = (\beta^{1\prime}, \beta^{2\prime}, \lambda')' \in \Gamma \subset E^{3k}$ is the unknown vector of amplitudes and frequencies. This NRM can be used to model random processes in which a quasiperiodic character with unknown frequencies can be expected.

Another type of mean value m_γ, nonlinearly depending on parameters γ, is the function

$$m_\gamma(t) = \gamma_1 e^{\gamma_2 t}; t \in T,$$

where usually $\gamma = (\gamma_1, \gamma_2)' \in \Gamma = (0, \infty) \times (-\infty, 0)$. This mean value is called the *exponential trend* and can be used, for example, in medicine, for the description of a decrease amount of medicine injected into a patient. The parameter γ_1 represents an amount of medicine injected and γ_2 characterizes the ability of a patient's body to absorb the medicine. This mean value can also be written in the form

$$m_\gamma(t) = \gamma_1 \gamma_2^t; t \in T; \gamma = (\gamma_1, \gamma_2)' \in \Gamma = (0, \infty) \times (0, 1).$$

This exponential trend can be changed to the *modified exponential trend* by adding one new parameter. This is defined by

$$m_\gamma(t) = \gamma_1 + \gamma_2 \gamma_3^t; t \in T; \gamma = (\gamma_1, \gamma_2, \gamma_3)' \in \Gamma = (-\infty, \infty) \times (0, \infty) \times (0, 1).$$

Another example of an NRM is a *logistic trend*. The mean value of this model is given by

$$m_\gamma(t) = \frac{\gamma_1}{1 + \gamma_2 \gamma_3^t}; t \in T, \gamma = (\gamma_1, \gamma_2, \gamma_3)' \in \Gamma = (-\infty, \infty) \times (0, \infty) \times (0, 1).$$

This function can be used for the modeling for the sale of new products, such as new cars in some region. It belongs to a *family of s-curves* symmetric about an rigorous point. This point, for the logistic trend, is at $t_{in} = -\ln \gamma_2 / \ln \gamma_3$.

For some economic applications the following NRM can be a good model for some random processes. Let

$$X(t) = m_\gamma(t) + \varepsilon(t); t \in T, \gamma \in \Gamma,$$

where

$$m_\gamma(t) = \sum_{i=1}^{k} \beta_i^1 f_i(t) + \varepsilon(t) \text{ for } t \in T_1,$$

$$m_\gamma(t) = \sum_{i=1}^{k} \psi_{i,\gamma}(t) f_i(t) + \varepsilon(t) \text{ for } t \in T_t,$$

and

$$m_\gamma(t) = \sum_{i=1}^{k} \beta_i^2 f_i(t) + \varepsilon(t) \text{ for } t \in T_2,$$

where $T_1 = (0, t_1)$, $T_t = (t_1, t_2)$, and $T_2 = (t_2, t_3)$.

The functions $\psi_{i,\gamma}(.)$ are defined by

$$\psi_{i,\gamma}(t) = \left[1 - g_\gamma \left(1 - \frac{t - t_1}{t_2 - t_1} \right) \right] \beta_i^1 + g_\gamma \left(\frac{t - t_1}{t_2 - t_1} \right) \beta_i^2; t \in T_t, i = 1, 2, ..., k,$$

where $g_\gamma(0) = 0, g_\gamma(1) = 1$, and

$$g_\gamma(t) = \frac{1}{1 + \exp\{\gamma_1 + \gamma_2 \cot(\pi t)\}}; t \in (0, 1).$$

For the parameters γ of this model we assume that

$$\gamma = (\gamma_1, \gamma_2)' \in \Gamma = E^1 \times (0, \infty).$$

This NRM can be called *seasonal*, it has regression parameters β^1 in the *season* T_1, β^2 in the *season* T_2, and there is a continuous change of parameter β^1 to parameter β^2 during the *transient season* T_t. This change is realized by using a function $g_\gamma(.)$ which depends on the parameter γ nonlinearly.

There are also *nonparametric models* for mean values of random processes, but we shall not deal with these models in this book.

Statistical models can be used not only for mean values, but also for covariance functions of random processes. We now give some models which are satisfactory from the point of view of applications.

If the covariance function $R(.,.)$ of X can be written in the form

$$R(s, t) = r(|s - t|); s, t \in T,$$

where $r(.)$ is a suitable function fulfilling the conditions given below, then X is called the *covariance stationary* and $r(.)$ is called the covariance, or *autocovariance, function* of X. It is common that the autocovariance function $r(.)$ is usually also denoted by $R(.)$, but it is now a function of one variable, the absolute difference of time periods. An interpretation of the covariance stationarity is that a linear dependence between random variables which constitute such a random process is the same for all random variables which have the same time distance. Thus linear dependence between $X(s)$ and $X(t)$ depends only on $|s - t|$ and is the same as the linear dependence between $X(s + u)$ and $X(t + u)$ for any $u \in T$. This can also be expressed by the property

$$R(t) = Cov(X(s); X(s + t))$$

which is true for all s and any fixed t.

We remark that the model of covariance stationarity is a *nonparametric model* on the covariance function of X.

For the covariance function $R(.)$ of the covariance stationary random process X we have

$$D[X(t)] = R(0),$$
$$R(-t) = R(t),$$
$$|R(t)| \leq R(0) \text{ for all } t \in T,$$

and

$$\sum_{i=1}^{n}\sum_{j=1}^{n} a_i a_j R(|t_i - t_j|) \geq 0$$

for all $a \in E^n$, all $t_1, ..., t_n \in T$, and every integer n.

A random process X which is stationary in mean value and in covariance function will be called *stationary*.

Covariance functions $R(.)$ of covariance stationary processes can depend on a covariance parameter ν and thus we can write

$$R(.) \in \mathcal{R} = \{R_\nu(t); t \in T, \nu \in \Upsilon\},$$

where Υ is a parametric space.

Example 2.2.1. Let us consider a random process X defined by

$$X(t) = \sum_{j=1}^{l}(X_j \cos jt + Y_j \sin jt); t \in T,$$

where it is assumed that X_j and Y_j are independent random variables with $E[X_j] = E[Y_j] = 0$ and $D[X_j] = D[Y_j] = \sigma_j^2 \in (0, \infty)$ for all $j = 1, 2, ..., l$. Then X is a centered random process with covariance functions

$$R_\nu(s, t) = \sum_{j=1}^{l} \sigma_j^2(\cos js \cos jt + \sin js \sin jt)$$

$$= R_\nu(|s - t|) = \sum_{j=1}^{l} \sigma_j^2 \cos j(s - t); s, t \in T,$$

$$\nu = (\sigma_1^2, ..., \sigma_l^2)' \in \Upsilon = (0, \infty)^l.$$

We see that this random process is covariance stationary and that this autocovariance function $R_\nu(.)$ depends on the parameter ν linearly.

Example 2.2.2. Another example of the linearly dependent covariance function $R_\nu(.,.)$ on parameter ν is the covariance function of a random process X defined by

$$X(t) = \sum_{j=1}^{l} X_j v_j(t); t \in T,$$

where X_j are independent random variables with $E[X_j] = 0$, and variances $D[X_j] = \sigma_j^2 \in (0, \infty)$ for all $j = 1, 2, ..., l$ and $v_j(.)$ are known given functions. Then X is a centered random process with covariance function

$$R_\nu(s, t) = \sum_{j=1}^{l} \sigma_j^2 v_j(s) v_j(t); s, t \in T, \nu = (\sigma_1^2, ..., \sigma_l^2)' \in \Upsilon = (0, \infty)^l.$$

Example 2.2.3. Let us write the following additive model for a random process X :

$$X(t) = m_\beta(t) + \varepsilon(t) = \sum_{i=1}^{k} \beta_i f_i(t) + \sum_{j=1}^{l} \varepsilon_j v_j(t); t \in T, \beta \in E^k,$$

where $f_1(.), ..., f_k(.)$ and $v_1(.), ..., v_l(.)$ are given known functions, $f_i(.) \neq v_j(.)$ for all i, j. Let $\varepsilon = (\varepsilon_1, ..., \varepsilon_l)'$ be a random vector with $E[\varepsilon] = 0$ and with a covariance matrix Σ^ε. Then

$$E_\beta[X(t)] = \sum_{i=1}^{k} \beta_i f_i(t); t \in T, \beta \in E^k$$

and

$$R_{\Sigma^\varepsilon}(s, t) = Cov(X(s); X(t)) = \sum_{i=1}^{l} \sum_{j=1}^{l} \Sigma_{ij}^\varepsilon v_i(s) v_j(t); s, t \in T.$$

This model, for the covariance functions $R_{\Sigma^\varepsilon}(., .)$, is an LRM with unknown parameters Σ^ε. It is a linear combination of known functions $v_i(.) v_j(.)$ defined on $T \times T$ with unknown parameters $\Sigma_{ij}^\varepsilon; i, j = 1, 2, ..., l$.

In a special case when ε_j are uncorrected with $D[\varepsilon_j] = \sigma_j^2; j = 1, 2, ..., l$ we can write

$$R_\nu(s, t) = \sum_{j=1}^{l} \sigma_j^2 v_j(s) v_j(t); s, t \in T, \nu = (\sigma_1^2, ..., \sigma_l^2)' \in \Upsilon = (0, \infty)^l.$$

These models for a random process X will be called *double linear regression models* (DOLRMs) with regression parameter β and variance-covariance parameter Σ^ε or ν. The reason for this notation is that we have LRMs for both the mean value, as well as for the covariance function, of the random process X.

As an example of autocovariance functions $R_\nu(.)$, which depend on parameter ν nonlinearly are the covariance functions

$$R_\nu(t) = \sigma^2 \exp\{-\alpha t\}; t \in T, \nu = (\sigma^2, \alpha)' \in \Upsilon = (0, \infty) \times (0, \infty).$$

Another such example is

$$R_\nu(t) = \sigma^2 \exp\{-\alpha t\} \cos \beta t; t \in T, \nu = (\sigma^2, \alpha, \beta)' \in \Upsilon = (0, \infty) \times (0, \infty) \times \Lambda$$

where $\Lambda = (-\infty, \infty)$ for a random process with continuous time and where $\Lambda = \langle -\pi, \pi \rangle$ for a time series.

An NRM

$$X(t) = m_\gamma(t) + \varepsilon(t); t \in T, \gamma \in \Gamma,$$

with covariance functions

$$R_\nu(s, t) = Cov(X(s); X(t)); s, t \in T, \nu \in \Upsilon,$$

or with autocovariance functions

$$R_\nu(t) = Cov(X(s); X(s + t)); t \in T, \nu \in \Upsilon,$$

where the dependence of R_ν on ν is also nonlinear, will be called a *double nonlinear regression model* (DONRM).

We can use for random processes, regression models which are linear for mean values and nonlinear for covariance functions and vice versa. These models can be called a *linear-nonlinear regression model* (L-NRM) or a *nonlinear-linear regression model* (N-LRM).

2.3 Spectral Theory

We start this section of the book with an example.

Example 2.3.1. Let us consider a random process X of a similar form as that in Example 2.2.1 given by

$$X(t) = \sum_{j=1}^{l} (X_j \cos \lambda_j t + Y_j \sin \lambda_j t); t \in T, \lambda \in \Lambda,$$

where it is assumed that X_j and Y_j are random variables with mean values $E[X_j] = E[Y_j] = 0$ for all $j = 1, 2, ..., l$, and $\lambda_j \in \Lambda; j = 1, 2, ..., l$, where $\Lambda = (-\infty, \infty)$ for a random process with continuous time and $\Lambda = \langle -\pi, \pi \rangle$ for a time series.

Using the equalities

$$\cos \alpha = \frac{1}{2}(e^{i\alpha} + e^{-i\alpha}), \sin \alpha = \frac{1}{2i}(e^{i\alpha} - e^{-i\alpha}),$$

where i denotes the complex unity, we can write the random process X in the form

$$X(t) = \sum_{j=-l,j\neq 0}^{l} Z_j e^{i\lambda_j t},$$

and where $\lambda_{-j} = -\lambda_j$ and Z_j are complex-valued random variables,

$$Z_j = \frac{1}{2}(X_j - iY_j) \text{ for a positive } j$$

and

$$Z_j = \frac{1}{2}(X_j + iY_j) \text{ for a negative } j.$$

It is easy to show that the covariance function $R(.,.)$ of X is given by

$$R(s,t) = E[X(s)\bar{X}(t)] = E\left[\sum_{j=-l,j\neq 0}^{l} Z_j e^{i\lambda_j s} \sum_{k=-l,k\neq 0}^{l} \bar{Z}_k e^{-i\lambda_k t}\right]$$

$$= \sum_{j=-l,j\neq 0}^{l} \sum_{k=-l,k\neq 0}^{l} E[Z_j \bar{Z}_k] e^{i(\lambda_j s - \lambda_k t)}; s,t \in T,$$

and we get that $R(s,t) = R(|s-t|)$ for all $s,t \in T$, and the process X is covariance stationary, if and only if the random variables Z_j are uncorrelated,

$$E[Z_j \bar{Z}_k] = 0 \text{ for } j \neq k.$$

In this case we can write

$$R(t) = \sum_{j=-l,j\neq 0}^{l} D[Z_j] e^{i\lambda_j t} = 2\sum_{j=1}^{l} D[Z_j] \cos \lambda_j t,$$

for all t, where

$$2D[Z_j] = \frac{1}{2}(D[X_j] + D[Y_j]); j = 1,2,...,l.$$

Next it is possible to write

$$R(t) = \sum_{j=-l,j\neq 0}^{l} D[Z_j] e^{i\lambda_j t} = \int_\Lambda e^{i\lambda t} dF(\lambda) \text{ for all } t,$$

where the region of integration Λ is $(-\infty, \infty)$ for a random process with continuous time and where $\Lambda = \langle -\pi, \pi \rangle$ for a time series, and

$$F(\lambda) = \sum_{\{j : \lambda_j < \lambda\}} D[Z_j]; \lambda \in \Lambda.$$

The function $F(.)$ is called the spectral distribution function of X and the last expression for $R(.)$ is called the spectral decomposition of autocovariance function $R(.)$.

By analogy we can write

$$X(t) = \sum_{j=-l, j \neq 0}^{l} Z_j e^{i\lambda_j t} = \int_{\Lambda} e^{i\lambda t} dZ(\lambda); t \in T,$$

where $Z(.)$ is an *orthogonal random measure* defined on a Borel set of E^1 by

$$Z(\{\lambda_j\}) = Z_j; j = 1, 2, ..., l \text{ and } Z(B) = 0 \text{ if } \lambda_j \notin B; j = 1, 2, ..., l.$$

The last expression for X, through a stochastic integral with respect to an orthogonal random measure Z, is called the spectral decomposition of X. It can be said that the random process X studied in this example has the *finite discrete spectrum* with support at points $\lambda_1, ..., \lambda_l \in \Lambda$.

The *spectral theory of centered stationary random processes* says that every such random process X has the *spectral decomposition*

$$X(t) = \int_{\Lambda} e^{i\lambda t} dZ(\lambda); t \in T,$$

with respect to some *orthogonal random measure* Z defined on Borel sets of the real line. The autocovariance function $R(.)$ of X has the *spectral decomposition*

$$R(t) = \int_{\Lambda} e^{i\lambda t} dF(\lambda) \text{ for all } t,$$

where the region of integration Λ is equal to $(-\infty, \infty)$ for a random process with continuous time and $\Lambda = \langle -\pi, \pi \rangle$ for a time series. F is the *spectral distribution function* of X defined on E^1. The random measure Z and the spectral distribution function F are connected through the relation

$$E[Z(B)\bar{Z}(C)] = \int_{B \cap C} dF(\lambda)$$

from which we get the orthogonality of Z, this means,

$$E[Z(B)\bar{Z}(C)] = 0 \text{ if } B \cap C \text{ is the empty set}$$

and

$$E[|Z(B)|^2] = \int_B dF(\lambda) \text{ for any Borel set } B.$$

The interpretation of the spectral theory is that every centered stationary random process is a mixture of goniometric functions with different frequencies and with random amplitudes. The dispersion of an amplitude by the given frequency depends on the increase of the spectral distribution function at the given frequency.

In many practical applications of theory of random processes we assume that there exists a *spectral density* $f(\lambda); \lambda \in \Lambda$, of a random process X. The spectral density $f(.)$ is defined by

$$f(\lambda) = \frac{dF(\lambda)}{d\lambda}; \lambda \in \Lambda.$$

This spectral density exists if the covariance function $R(.)$ of X is summable that is, if

$$\int_0^\infty |R(t)| \, dt, \text{ or } \sum_{t=0}^\infty |R(t)| \text{ is finite.}$$

Then the spectral density $f(.)$ can be computed according to the relations

$$f(\lambda) = \frac{1}{2\pi} \int_{-\infty}^\infty R(t)e^{i\lambda t} dt = \frac{1}{2\pi} \int_{-\infty}^\infty R(t) \cos \lambda t dt; \lambda \in (-\infty, \infty),$$

if $T = (-\infty, \infty)$, or

$$f(\lambda) = \frac{1}{2\pi} \sum_{t=-\infty}^\infty R(t)e^{i\lambda t} = \frac{1}{2\pi} \sum_{t=-\infty}^\infty R(t) \cos \lambda t; \lambda \in \Lambda = \langle -\pi, \pi \rangle,$$

for $T = \{..., -2, -1, 0, 1, 2, ...\}$.

Conversely, the covariance function $R(.)$ of X can be computed from the spectral density using the relation

$$R(t) = \int_\Lambda \cos \lambda t f(\lambda) d\lambda; t \in T,$$

where, as usual, $\Lambda = (-\infty, \infty)$ for $T = (-\infty, \infty)$ and $\Lambda = \langle -\pi, \pi \rangle$ for $T = \{..., -2, -1, 0, 1, 2, ...\}$.

It is well known that spectral densities are symmetric nonnegative integrable functions

$$f(\lambda) \geq 0, f(-\lambda) = f(\lambda); \lambda \in \Lambda \text{ and } \int_\Lambda f(\lambda)d\lambda = R(0).$$

They can, in the same way as covariance functions, depend on some parameter ν. Thus we have parametric families $f_\nu; \nu \in \Upsilon$, of spectral densities.

Example 2.3.2. Let us consider a random process X with the covariance functions

$$R_\nu(t) = \sigma^2 \exp\{-\alpha t\}; t \in (-\infty, \infty), \nu = (\sigma^2, \alpha)' \in \Upsilon = (0, \infty) \times (0, \infty).$$

Then the spectral densities $f_\nu(\lambda); \lambda \in (-\infty, \infty), \nu \in \Upsilon$, are

$$f_\nu(\lambda) = \frac{1}{2\pi} \int_{-\infty}^{\infty} R(t)e^{i\lambda t}dt = \frac{1}{2\pi} \int_{-\infty}^{\infty} \sigma^2 \exp\{-\alpha t\}e^{i\lambda t}dt = \frac{\sigma^2}{\pi} \frac{\alpha}{\alpha^2 + \lambda^2}.$$

For any $\nu = (\sigma^2, \alpha, \beta)' \in \Upsilon = (0, \infty) \times (0, \infty) \times (-\infty, \infty)$ we have

$$R_\nu(t) = \sigma^2 \exp\{-\alpha t\} \cos \beta t; t \in (-\infty, \infty),$$

and we get the spectral densities

$$f_\nu(\lambda) = \frac{\sigma^2 \alpha}{\pi} \frac{\lambda^2 + \alpha^2 + \beta^2}{\lambda^4 + 2(\alpha^2 - \beta^2)^2 \lambda^2 + (\alpha^2 + \beta^2)^2}; \lambda \in (-\infty, \infty), \nu \in \Upsilon.$$

2.4   Models for Time Series

Since time series are special types of random processes, all models already involved in the preceding sections of this book are also valid for time series. But there are also models which have meaning only in the case when time is discrete, that is when $T = Z = \{..., -2, -1, 0, 1, 2, ...\}$, the set of all integers. We shall give briefly such models, first in the case when the mean value of time series X is equal to zero, or when X is the centered time series. The following results can be found in Box and Jenkins (1976) and Brockwell and Davis (1987). A basic model is white noise. The time series $Y = \{Y(t); t \in T\}$ is called *white noise* if $Y(t); t \in T$, are independent random variables with $E[Y(t)] = 0$ and with the same variance $D[Y(t)] = \sigma^2$. White noise Y is called *Gaussian*, if $Y(t); t \in T$, have the Gaussian distribution. White noise is a base for the construction of other types of time series. It is clear that white noise is a stationary centered time series with covariance functions $R_\nu(0) = \sigma^2; \nu = \sigma^2 \in \Upsilon = (0, \infty)$, and $R_\nu(t) = 0$ for $t \neq 0$. The spectral densities $f_\nu(.)$ are constant functions

$$f_\nu(\lambda) = \frac{\sigma^2}{2\pi}; \lambda \in \Lambda = \langle 0, \pi \rangle, \nu \in \Upsilon.$$

A *moving average of order q* (MA(q)) time series X is defined as a linear combination of a white noise Y by the equation

$$X(t) = Y(t) + \sum_{j=1}^{q} b_j Y(t-j); t \in T,$$

where $b = (b_1, ..., b_q)' \in E^q$. It is possible to show that the MA(q) are centered stationary time series with covariance functions

$$R_b(t) = \sigma^2 \sum_{j=0}^{q-t} b_j b_{j+t} \text{ for } t = 0, 1, ..., q,$$

$$R_b(t) = 0 \text{ for } t \geq q+1; b_0 = 1, b \in E^q.$$

The last property, $R_b(t) = 0$ for $t \geq q+1$, of the covariance function is a characteristic property of the MA(q) time series. The spectral densities $f_b(.)$ of the MA(q) X are

$$f_b(\lambda) = \frac{\sigma^2}{2\pi} \left[\left(1 + \sum_{j=1}^{q} b_j \cos j\lambda \right)^2 + \left(\sum_{j=1}^{q} b_j \sin j\lambda \right)^2 \right]; \lambda \in \Lambda = \langle 0, \pi \rangle.$$

The notion of MA(q) time series can be generalized to an infinite MA(∞) time series. Let $b = \{b_j\}_{j=-\infty}^{\infty}$ be an *absolutely summable* doubly infinite sequence, that is there exists the finite limit

$$\lim_{n \to \infty} \sum_{j=-n}^{n} |b_j|.$$

Then it is shown in Fuller (1976) that there exists a time series $X(t); t \in T$, such that

$$\lim_{n \to \infty} E \left[X(t) - \sum_{j=-n}^{n} b_j Y(t-j) \right]^2 = 0.$$

We shall call this time series $X(t); t \in T$, an *infinite MA(∞) time series* and we shall write

$$X(t) = \sum_{j=-\infty}^{\infty} b_j Y(t-j); t \in T.$$

MA(∞) time series $X(t); t \in T$, has mean a value function equal zero and a covariance function

$$R_b(t) = \sigma^2 \sum_{j=-\infty}^{\infty} b_j b_{j+t}; t = 0, 1, ...$$

We remark that in some cases $b_j = 0$ for $j = -1, -2, ...$

An *autoregressive time series of order p (AR(p))* time series X is defined as a linear combination of the past of X and of a white noise term $Y(t)$:

$$X(t) = \sum_{j=1}^{p} a_j X(t-j) + Y(t); t \in T,$$

where $a = (a_1, ..., a_p)' \in E^p$. The AR(p) time series X is a centered time series. Its second-order properties depend on the vector a which defines such a time series.

Let

$$\phi_a(x) = 1 - \sum_{j=1}^{p} a_j x^j; x \in E^1,$$

be the polynomial determined by the vector a. Then the following well-known theorem gives a characterization of the AR(p) time series X.

Theorem 2.4.1. *The AR(p) time series X, defined by*

$$X(t) = \sum_{j=1}^{p} a_j X(t-j) + Y(t); t \in T,$$

where $a = (a_1, ..., a_p)' \in E^p$ is stationary if and only if all the roots of the equation $\phi_a(x) = 0$ have modulus greater than one.

If X is stationary then the covariance function $R_a(.)$ of X can be computed using the *Yule-Walker equations*

$$R_a(t) = \sum_{j=1}^{p} a_j R_a(t-j); t = 1, 2, ...$$

The spectral densities $f_a(.); a \in (-1, 1)$, of the stationary AR(p) time series X are given by

$$f_a(\lambda) = \frac{\sigma^2}{2\pi} \left[\left(1 - \sum_{j=1}^{p} a_j \cos j\lambda \right)^2 + \left(\sum_{j=1}^{p} a_j \sin j\lambda \right)^2 \right]^{-1}; \lambda \in \Lambda = \langle 0, \pi \rangle.$$

Example 2.4.1. Let X be the AR(1) with parameter $a = \rho$ given by the equation

$$X(t) = \rho X(t-1) + Y(t); t \in T.$$

Then it is well known that X is stationary iff $\rho \in (-1, 1)$. In this case

$$X(t) = \sum_{j=0}^{\infty} \rho^j Y(t-j); t \in T,$$

and

$$R_\nu(t) = \frac{\sigma^2}{1 - \rho^2} \rho^t; t = 0, 1, ..., \quad \nu = (\sigma^2, \rho)' \in \Upsilon = (0, \infty) \times (-1, 1).$$

The spectral densities $f_\nu(.)$ are given by

$$f_\nu(\lambda) = \frac{\sigma^2}{2\pi} \frac{1}{1 - 2\rho \cos \lambda + \rho^2}; \lambda \in \Lambda = \langle 0, \pi \rangle, \nu \in \Upsilon.$$

For $|\rho| \geq 1$ the AR(p) time series X has an explosive character and it is not stationary.

The centered time series X, defined by the equation

$$X(t) = \sum_{j=1}^{p} a_j X(t-j) + Y(t) + \sum_{j=1}^{q} b_j Y(t-j); t \in T,$$

where $a = (a_1, ..., a_p)' \in E^p$ and $b = (b_1, ..., b_q)' \in E^q$, is called the *autoregressive moving average time series of order p, q (ARMA(p,q))*.

Theorem 2.4.2. *The ARMA(p,q) time series is stationary, iff the roots of the polynomial $\phi_a(x) = 1 - \sum_{j=1}^{p} a_j x^j; x \in E^1$, have modulus grater than one.*

For the stationary ARMA(p,q) we can find their spectral densities for suitable $a \in E^p$ and for any $b \in E^q$ according to the equation

$$f_{a,b}(\lambda) = \frac{\sigma^2}{2\pi} \frac{\left(1 + \sum_{j=1}^{q} b_j \cos j\lambda\right)^2 + \left(\sum_{j=1}^{q} b_j \sin j\lambda\right)^2}{\left(1 - \sum_{j=1}^{p} a_j \cos j\lambda\right)^2 + \left(\sum_{j=1}^{p} a_j \sin j\lambda\right)^2}; \lambda \in \Lambda = \langle 0, \pi \rangle.$$

Example 2.4.2. Let us consider the ARMA($1,1$) time series X with parameters $a = \rho$ and b. Then X is stationary iff $\rho \in (-1, 1)$. In this case it can be shown that the covariance functions $R_\nu(.)$, with parameter $\nu = (\sigma^2, \rho, b)'$ belonging to $\Upsilon = (0, \infty) \times (-1, 1) \times (-\infty, \infty)$ of X, are given by

$$R_\nu(0) = \sigma^2 \left[1 + \frac{(\rho + b)^2}{1 - \rho^2} \right]$$

and

$$R_\nu(t) = \sigma^2 \left[(\rho + b)\rho^{t-1} + \frac{(\rho + b)^2}{1 - \rho^2} \rho^t \right] \text{ for } t \geq 1.$$

It should be noted that for $t \geq 2$, $R(t) = \rho R(t - 1)$. This is reminiscent of the exponentially decaying autocovariance function of an AR(1) time series. The distinction is that for the ARMA(1,1) time series $R_\nu(1) \neq \rho\sigma^2/(1-\rho^2)$. The spectral densities of the stationary ARMA(1,1) are

$$f_\nu(\lambda) = \frac{\sigma^2}{2\pi} \frac{1 + 2b\cos\lambda + b^2}{1 - 2\rho\cos\lambda + \rho^2}; \lambda \in \Lambda = \langle 0, \pi \rangle, \nu = (\sigma^2, \rho, b)' \in \Upsilon.$$

The usefulness of an ARMA time series stems from the ability of the covariance function, or of the spectral density, to take a wide variety of shapes without requiring either p or q to be particularly large. An immediate consequence is that the second-order properties of a stationary time series can often be well approximated by an ARMA time series which is economical in its use of parameters.

All the models we have already mentioned are based on the assumption that the time series have mean values equal to zero. Now we give a class of models where this condition is omitted. It is the class of *autoregressive integrated moving average (ARIMA)* models. These models serve as good models for many time series whose mean value is not stationary and can be assumed to follow a polynomial trend. They are based on *differencing time series* which is a simple device for removing the polynomial trend from the time series.

The *first difference* ∇X of X is defined by

$$\nabla X(t) = X(t) - X(t - 1); t \in T,$$

and the *higher-order differences* by

$$\nabla^k X(t) = \nabla(\nabla^{k-1} X(t)); t \in T, k = 2, 3, \ldots.$$

Thus, for example,

$$\nabla^2 X(t) = \nabla(\nabla X(t)) = \nabla(X(t) - X(t-1)) = X(t) - 2X(t-1) + X(t-2).$$

Differencing is a simple device for removing polynomial trends from time series. It can be shown that if the mean value of X is a polynomial of degree k, then the mean value of $\nabla^k X$ is a constant and thus $\nabla^k X$ is stationary in mean value.

Example 2.4.3. Let

$$X(t) = \beta_1 + \beta_2 t + \varepsilon(t); t \in T.$$

Then

$$\nabla X(t) = \beta_1 + \beta_2 t + \varepsilon(t) - \beta_1 + \beta_2(t-1) + \varepsilon(t) - \varepsilon(t-1) = \beta_2 + \nabla \varepsilon(t); t \in T,$$

and we see that ∇X is stationary in mean value. For the covariance function $R^\nabla(.,.)$ of ∇X we get

$$R^\nabla(s,t) = R(s,t) - R(s-1,t) - R(s,t-1) + R(s-1,t-1); s,t \in T,$$

where $R(.,.)$ is the covariance function of X. If X is covariance stationary, then

$$R^\nabla(t) = 2R(t) - R(t-1) - R(t+1)$$

and we see that ∇X is also covariance stationary. The variance of ∇X is equal to $R^\nabla(0) = D[\nabla X(t)] = 2(R(0) - R(1))$ and is smaller than $D[X(t)]$ if $R(1)$ is a positive number. If $R(1)$ is negative, then the variance of ∇X is greater than the variance of X.

We say that X is an ARIMA *time series of order* p,d,q, if the dth difference $\nabla^d X$ of X is a stationary ARMA(p,q). These time series are called ARIMA(p,d,q) time series.

The differences ∇^d can be computed using the back-shift operator B defined by

$$B(X(t)) = X(t-1); t \in T,$$

from which we get

$$\nabla^d X = (1-B)^d X,$$

where $B^k X = B(B^{k-1}X)$ for $k \geq 2$ and thus $B^k X(t) = X(t-k); t \in T.$

If we denote by ϕ the polynomial defined by

$$\phi(B) = 1 - \sum_{j=1}^{p} a_j B^j$$

and by ψ the polynomial defined by

$$\psi(B) = 1 - \sum_{j=1}^{q} b_j B^j,$$

then the ARIMA(p,d,q) time series X can be written by the equality

$$\phi(B)(1 - B)^d X = \psi(B)Y.$$

Example 2.4.4. The time series

$$X(t) = X(t - 1) + Y(t); t \in T,$$

is a simple example of an ARIMA($0,1,0$) time series, while the time series X defined by

$$(1 - 0.3B)(1 - B)X = (1 + 3B)Y,$$

or

$$X(t) = 1.3X(t - 1) - 0.3X(t - 2) + Y(t) + 3Y(t - 1)$$

is an example of an ARIMA($1,1,1$) model for time series. We remark that this time series can also be considered as the nonstationary ARMA($2,1$) with roots of the characteristic polynomial equal to 0.3 and 1.

3

Estimation of Time Series Parameters

3.1 Introduction

In the preceding chapters we have described some parametric models for random processes and time series. In all the introduced parametric models there are parameters β or γ of mean values and parameters ν of covariance functions which are unknown in practical applications and which should be estimated from the random process, or time series, data. By this data we mean a real vector x of realizations of a finite observation $X_O = \{X(t); t \in T_O\}$ of a random process $X(.) = \{X(t); t \in T\}$. Usually $X_O = (X(1), ..., X(n))'$ if $X(.)$ is a time series and $X_O = (X(t_1), ..., X(t_n))'$ if X_O is a discrete observation of the random process $X(.)$ with continuous time at time points $t_1, ..., t_n$. The *length of observation* n is some natural number. In this chapter we shall assume that $t_{i+1} - t_i = d; i = 1, 2, ..., n-1$, that is we have an observation X_O of $X(.)$ at *equidistant time points* $t_1, ..., t_n \in T$. Next we shall omit the subscript O and we shall denote the finite observation of the length n of a time series or of a random process $X(.)$ by the unique notation

$$X = (X(1), ..., X(n))' \text{ or by } X_n = (X(1), ..., X(n))'$$

to denote its dependence on n. The vector X will be, in both cases, called the *finite time series observation*. The vector $x = (x(1), ..., x(n))'$ where $x(t)$ is a realization of $X(t); t = 1, 2, ..., n$ will be called the *time series data*.

In this chapter we shall consider mainly the additive model for time series $X(.)$ given by

$$X(t) = m(t) + \varepsilon(t); t \in T, m(.) \in \mathcal{M}.$$

Then for the finite observation $X = (X(1), ..., X(n))'$ of the length n we get

$$X = m + \varepsilon, m \in \mathcal{M},$$

where $m = (m(1), ..., m(n))' = E[X]$ is the mean value of X and the $n \times 1$

vector $\varepsilon = (\varepsilon(1), ..., \varepsilon(n))'$ is the random vector of errors of the model with mean value $E[\varepsilon] = 0$ and with $Cov(\varepsilon) = Cov(X) = \Sigma$. If we denote, as usual, by $R(.,.)$ the covariance function of $X(.)$, then we have

$$\Sigma_{st} = R(s,t); s, t = 1, 2, ..., n.$$

Thus the covariance matrix Σ of X is determined by the covariance function $R(.,.)$ of the time series $X(.)$.

If we have no model on the mean value $m(.)$ of $X(.)$, that is if we assume that $\mathcal{M} = E^n$, then the natural estimator $\tilde{m}$ of m is given by

$$\tilde{m}(t) = X(t); t = 1, 2, ..., n.$$

The estimator $\tilde{m}$ is an unbiased, but inconsistent estimator of m with possibly large covariance matrix Σ. The residuals $\tilde{\varepsilon}$ for this estimator are given by

$$\tilde{\varepsilon}(t) = X(t) - \tilde{m}(t) = 0; t = 1, 2, ..., n,$$

and thus we have no information about the unknown covariance function $R(.,.)$ of $\varepsilon(.)$. It follows that if we have no model on the mean value $m(.)$ of the time series $X(.)$, then it is not possible to estimate the covariance function $R(.,.)$ on the base of one observation X of this time series.

In many practical problems of time series analysis we do not know either the mean value $m(.)$, or the covariance function $R(.,.)$ of $X(.)$ and only the realization $x = (x(1), ..., x(n))'$ of the finite observation X of $X(.)$ is available. Thus if we want to estimate both the mean value $m(.)$ and the covariance function $R(.,.)$ of $X(.)$ on the base of X, we must use some model for these unknown characteristics of $X(.)$.

The basic models for the mean values and covariance functions of $X(.)$ have been described in Sections 2.2. and 2.4. We have shown that linear and nonlinear regression models for mean values and for covariance functions, together with the model of covariance stationarity, are the most frequent models used by modeling time series data. We shall study problems of the estimation of parameters of these models in the following sections of the book. Since finite observations of time series are random vectors, many results already derived in Chapter 1, are also valid for estimators in regression models with the error vector generated by some time series, that is with correlated errors.

3.2   Estimation of Mean Value Parameters

We have shown in the preceding chapters that the LRM for the mean value of time series is a very good model which suffices in many practical applications.

The LRM for time series $X(.)$ is given by

$$X(t) = \sum_{i=1}^{k} \beta_i f_i(t) + \varepsilon(t); t \in T, \beta \in E^k,$$

where $f_1, ..., f_k$ are given known functions. The time series $\varepsilon(.)$ is called *noise* and it is assumed that

$$\begin{aligned} E[\varepsilon(t)] &= 0, \\ Cov(\varepsilon(s); \varepsilon(t)) &= Cov(X(s); X(t)) = R(s,t); s, t \in T. \end{aligned}$$

Let $X = (X(1), ..., X(n))'$ be a finite observation of $X(.)$. Then the LRM for time series $X(.)$ generates the LRM for the observation X in the form

$$X = F\beta + \varepsilon, E[\varepsilon] = 0, \beta \in E^k, Cov(X) = \Sigma.$$

$F_{ij} = f_j(i); i = 1, 2, ..., n, j = 1, 2, ..., k$, are the elements of the $n \times k$ design matrix F and $\Sigma_{ij} = R(i,j); i, j = 1, 2, ..., n$, are the elements of the covariance matrix Σ. The basic problem of time series analysis in the LRM is to estimate the unknown parameter β of the model on the base of X. Since in practical applications the covariance function $R(.,.)$ of $X(.)$ is usually unknown, the OLSEs are those which can be used in these situations. We know that the OLSE $\hat{\beta}$ of β is given by

$$\hat{\beta}(X) = (F'F)^{-1}F'X$$

and has the covariance matrix given by

$$Cov_\Sigma(\hat{\beta}(X)) = (F'F)^{-1}F'\Sigma F(F'F)^{-1}.$$

As we have shown, the OLSE $\hat{\beta}$ is equal to the WELSE β^* of β if $\Sigma = \sigma^2 I_n$. But in other cases when the components of ε are not uncorrelated, the OLSE $\hat{\beta}$ is not equal to the WELSE β^*. As we shall show in the following theorem, the OLSE $\hat{\beta}$ also has good asymptotic properties for correlated errors for a large class of possible covariance functions $R(.,.)$ of observed time series $X(.)$.

Let X_n denote the finite observation of the time series $X(.)$ by the given n and let

$$\hat{\beta}(X_n) = (F'_n F_n)^{-1} F'_n X_n$$

be the OLSE of β computed at the observation X_n, where the dependence of F on n is denoted by writing $F = F_n$ and the dependence of Σ on n by $\Sigma = \Sigma_n$.

Let g be any vector from E^k, then we have

$$
\begin{aligned}
D[g'\hat{\beta}(X_n)] &= g'(F_n'F_n)^{-1}F_n'\Sigma_n F_n(F_n'F_n)^{-1}g \\
&= (\Sigma_n F_n(F_n'F_n)^{-1}g, F_n(F_n'F_n)^{-1}g),
\end{aligned}
$$

and using the Schwarz inequality

$$
\begin{aligned}
|(Aa, a)| &= \left| \sum_{i,j=1}^{n} A_{ij}a_i a_j \right| \\
&\le \left(\sum_{i,j=1}^{n} A_{ij}^2 \right)^{1/2} \left(\sum_{i,j=1}^{n} a_i^2 a_j^2 \right)^{1/2} = \|A\| \, \|a\|^2
\end{aligned}
$$

we get

$$
\begin{aligned}
D[g'\hat{\beta}(X_n)] &\le \|\Sigma\| \, \|F_n(F_n'F_n)^{-1}g\|^2 \\
&= \|\Sigma\| \, g'(F_n'F_n)^{-1}F_n'F_n(F_n'F_n)^{-1}g \\
&= \|\Sigma_n\| \, g'(F_n'F_n)^{-1}g.
\end{aligned}
$$

Since the value of k is fixed, the limit properties, for $n \to \infty$, of $D[g'\hat{\beta}(X_n)]$ depend on limit the properties of $\|\Sigma_n\|$ and of $(F_n'F_n)^{-1}$. The following idea describes a property which will be useful in this connection.

A sequence $\{A_n\}_{n=1}^{\infty}$ of $k \times k$ matrices is to be *of the order* $O(1/n)$ if it can be written as $A_n = 1/nG_n$, where G_n are $k \times k$ matrices having the nonzero limit $\lim_{n \to \infty} G_n = G$. The following theorem gives a sufficient condition for consistency of the OLSE $\hat{\beta}$.

Theorem 3.2.1. *Let, in the LRMs,*

$$
X_n = F_n\beta + \varepsilon_n, E[\varepsilon_n] = 0, Cov(X_n) = \Sigma_n, n = 1, 2, ...,
$$

the matrices $(F_n'F_n)^{-1}$ *are of the order* $O(1/n)$ *and let* $\lim_{n \to \infty} 1/n \|\Sigma_n\| = 0$. *Then the OLSE* $\hat{\beta}$ *is a consistent estimator of* β.

Proof. From the assumption of the theorem we can write $(F_n'F_n)^{-1} = 1/nG_n$ where $\lim_{n \to \infty} G_n = G$. Using this and the derived inequality for $D[g'\hat{\beta}_n]$ we get

$$
D[g'\hat{\beta}(X_n)] \le \frac{1}{n} \|\Sigma_n\| \, g'G_n g \to 0 \text{ as } n \to \infty.
$$

Consequence. *Let* $(F_n'F_n)^{-1}; n = 1, 2, ...,$ *be of the order* $O(1/n)$ *and let* $X(.)$ *be covariance stationary with an autocovariance function* $R(.)$ *having* *the property* $\lim_{t\to\infty} R(t) = 0$. *Then the OLSE $\hat{\beta}$ is the consistent estimator for β.*

Proof. The covariance matrix Σ_n of the observation X_n of a covariance stationary time series with the autocovariance function $R(.)$ is

$$\Sigma_n = \begin{pmatrix} R(0) & R(1) & \ldots & R(n-1) \\ R(1) & R(0) & \ldots & R(n-2) \\ . & . & \ldots & . \\ R(n-1) & R(n-2) & \ldots & R(0) \end{pmatrix}$$

and we can write

$$\frac{1}{n}\|\Sigma_n\| = \frac{1}{n}\left(\sum_{s=1}^{n}\sum_{t=1}^{n} R^2(s-t)\right)^{1/2}$$

$$= \frac{1}{n}\left[nR^2(0) + 2\sum_{t=1}^{n-1}(n-t)R^2(t)\right]^{1/2}$$

and thus, as $n \to \infty$,

$$\frac{1}{n}\|\Sigma_n\|\, g'G_n g = \left[\frac{R^2(0)}{n} + \frac{2}{n}\sum_{t=1}^{n}\left(1-\frac{t}{n}\right)R^2(t)\right]^{1/2} g'G_n g \to 0$$

for every $g \in E^k$, if $\lim_{t\to\infty} R(t) = 0$. The last statement follows from

$$0 \le \frac{2}{n}\sum_{t=1}^{n}\left(1-\frac{t}{n}\right)R^2(t) \le \frac{2}{n}\sum_{t=1}^{n} R^2(t) \to 0 \text{ for } n \to \infty,$$

which is a consequence of

$$\lim_{t\to\infty} R(t) = 0 \text{ iff } \lim_{t\to\infty} R^2(t) = 0 \text{ iff } \lim_{n\to\infty}\frac{1}{n}\sum_{t=1}^{n} R^2(t) = 0.$$

Remarks. 1. If the conditions of the theorem are fulfilled, then for any $g \in E^k$:

$$\lim_{n\to\infty} P(\left|g'\hat{\beta}(X_n) - g'\beta\right| > \varepsilon) = 0 \text{ for any } \varepsilon > 0$$

and thus $g'\hat{\beta}(X_n)$ converges in probability to $g'\beta$.

2. The condition $\lim_{t\to\infty} R(t) = 0$ is fulfilled for many models for autocovariance functions which have been introduced in the preceding chapters of this book, for example, for a covariance stationary ARMA time series. It can be assumed, by the practical applications of time series, that this condition on

autocovariance functions is fulfilled and thus the question of the consistency of OLSEs is reduced to the question as to whether $(F_n' F_n)^{-1}; n = 1, 2, ...,$ are of the order $O(1/n)$. We shall study this problem for the following examples.

3. The preceding statements are also valid for the WELSE β_R^* of β, since

$$D_R[\beta_R^*(X_n)] \le D_R[\hat{\beta}(X_n)]$$

for any covariance function $R(.,.)$, or any autocovariance function $R(.)$.

Example 3.2.1. Let us consider a mean value stationary time series $X(.)$:

$$X(t) = \beta + \varepsilon(t); t \in T, \beta \in E^1,$$

with an unknown mean value β. Then

$$X_n = F_n \beta + \varepsilon_n; \beta \in E^1,$$

where $F_n = j_n = (1, ..., 1)'$ and $(F_n' F_n)^{-1} = 1/n; n = 1, 2, ...$ We see that $(F_n' F_n)^{-1}; n = 1, 2, ...,$ are of the order $O(1/n)$ and thus the OLSE $\hat{\beta}$ of β is given by

$$\hat{\beta}(X_n) = \frac{1}{n} \sum_{t=1}^{n} X(t).$$

For the variance $D_R[\hat{\beta}(X_n)]$ we get

$$D_R[\hat{\beta}(X_n)] = \frac{1}{n^2} j_n' \Sigma_n j_n = \frac{1}{n^2} \sum_{s=1}^{n} \sum_{t=1}^{n} R(s, t).$$

If $X(.)$ is also covariance stationary, then

$$
\begin{aligned}
D_R[\hat{\beta}(X_n)] &= \frac{1}{n^2} \left(n R(0) + 2 \sum_{t=1}^{n-1} (n - t) R(t) \right) \\
&= \frac{R(0)}{n} + \frac{2}{n} \sum_{t=1}^{n} \left(1 - \frac{t}{n} \right) R(t)
\end{aligned}
$$

and $\lim_{n \to \infty} D_R[\hat{\beta}(X_n)] = 0$ if $\lim_{t \to \infty} R(t) = 0$.

In the case when $\varepsilon(.)$ is a white noise with the variance $R(0)$ we get $D[\hat{\beta}(X_n)] = R(0)/n$ and the question of a comparison of this variance with the variance $D_R[\hat{\beta}(X_n)]$ of $\hat{\beta}(X_n)$, by the assumption that $\varepsilon(.)$ has autocovariance function $R(.)$, arises. For simplicity, let us assume that $\varepsilon(.)$ is an AR(1) time series with the covariance function $R(t) = R(0)\rho^t; t =$

$0, 1, \ldots$. Thus the white noise and the AR(1) time series have the same variances equal to $R(0)$. Then we get

$$
\begin{aligned}
D_R[\hat{\beta}(X_n)] &= \frac{R(0)}{n} + \frac{2R(0)}{n^2} \sum_{t=1}^{n} (n-t)\rho^t \\
&= \frac{R(0)}{n} + \frac{2R(0)}{n^2} \frac{\rho^{n+1} + \rho[n(1-\rho) - 1]}{(1-\rho)^2}.
\end{aligned}
$$

It is easy to show that for $\rho < 0$:

$$
\frac{\rho^{n+1} + \rho[n(1-\rho) - 1]}{(1-\rho)^2} < 0 \text{ for all } n \geq 2,
$$

and thus the variance $D_R[\hat{\beta}(X_2)]$ by AR(1) error time series is smaller than the variance $D[\hat{\beta}(X_2)] = R(0)/2$ of the OLSE $\hat{\beta}$ computed by uncorrelated errors. The connection between these variances is just opposite for $\rho > 0$.

Example 3.2.2. Let us consider a time series $X(.)$ with a linear trend

$$
X(t) = \beta_1 + \beta_2 t + \varepsilon(t); t \in T; \beta = (\beta_1, \beta_2)' \in E^2.
$$

Then for the finite observations X_n we have the LRMs

$$
X_n = F_n\beta + \varepsilon_n; E[\varepsilon_n] = 0, \beta \in E^2, Cov(X_n) = \Sigma_n, n = 1, 2, \ldots
$$

with design matrices

$$
F_n' = \begin{pmatrix} 1 & 1 & \cdots & 1 \\ 1 & 2 & \cdots & n \end{pmatrix}.
$$

It is easy to show that

$$
F_n' F_n = \begin{pmatrix} n & [n(n+1)]/2 \\ [n(n+1)]/2 & [n(n+1)(2n+1)]/6 \end{pmatrix}
$$

and

$$
(F_n' F_n)^{-1} = \begin{pmatrix} [2(2n+1)]/[n(n-1)] & -6/[n(n-1)] \\ -6/[n(n-1)] & 12/[n(n^2-1)] \end{pmatrix}.
$$

Thus we get for the OLSE $\hat{\beta}$:

$$
\hat{\beta}(X_n) = (F_n' F_n)^{-1} F_n' X_n = (F_n' F_n)^{-1} \begin{pmatrix} \sum_{t=1}^{n} X(t) \\ \sum_{t=1}^{n} t X(t) \end{pmatrix}
$$

from which

$$\hat{\beta}_1(X_n) \;=\; [2(2n+1)]\,/\,[n(n-1)]\sum_{t=1}^{n} X(t) - 6/\,[n(n-1)]\sum_{t=1}^{n} tX(t),$$

$$\hat{\beta}_2(X_n) \;=\; -6/\,[n(n-1)]\sum_{t=1}^{n} X(t) + 12/\,[n(n^2-1)]\sum_{t=1}^{n} tX(t).$$

Other expressions for components of $\hat{\beta}(X_n)$ are

$$\hat{\beta}_1(X_n) = \bar{X}_n - \hat{\beta}_2(X_n)\bar{t}_n$$

and

$$\hat{\beta}_2(X_n) = \frac{\displaystyle\sum_{t=1}^{n}(X(t)-\bar{X}_n)(t-\bar{t}_n)}{\displaystyle\sum_{t=1}^{n}(t-\bar{t}_n)^2},$$

where $\bar{X}_n = 1/n\sum_{t=1}^{n} X(t)$ and $\bar{t}_n = 1/n\sum_{t=1}^{n} t = (n+1)/2$.

Next we can write

$$(F_n'F_n)^{-1} = \frac{1}{n}\left(\begin{array}{cc} [2(2n+1)]\,/\,[(n-1)] & -6/\,[(n-1)] \\ -6/\,[(n-1)] & 12/\,[(n^2-1)] \end{array}\right) = \frac{1}{n}G_n,$$

where

$$\lim_{n\to\infty} G_n = \left(\begin{array}{cc} 4 & 0 \\ 0 & 0 \end{array}\right)$$

and we see that $(F_n'F_n)^{-1}; n = 1, 2, ...$, are of the order $O(1/n)$. Thus the condition

$$\lim_{n\to\infty}\frac{1}{n}\left(\sum_{s=1}^{n}\sum_{t=1}^{n} R^2(s,t)\right)^{1/2} = 0,$$

or the condition

$$\lim_{t\to\infty} R(t) = 0$$

for covariance stationary time series, are sufficient for consistency of the OLSE $\hat{\beta}$ of parameter β of the linear trend of any time series $X(.)$. The last condition is fulfilled for $\varepsilon(.)$ by a white noise time series with the variance σ^2. Then

$$Var[\hat{\beta}(X_n)] = \sigma^2(F_n'F_n)^{-1}$$

and we see that $D[\hat{\beta}_1(X_n)]$ is of the order $O(1/n)$, but $D[\hat{\beta}_2(X_n)]$ is of the order $O(1/n^3)$ and thus converges to zero much quicker than $D[\hat{\beta}_1(X_n)]$ as $n \to \infty$. Next it can be seen that

$$Cov(\hat{\beta}_1(X_n); \hat{\beta}_2(X_n)) = -\frac{3}{2n+1}D[\hat{\beta}_1(X_n)] < 0.$$

The estimated mean value $\hat{m}(.)$ defined by

$$\hat{m}(t) = \hat{\beta}_1 + \hat{\beta}_2 t; t = 1, 2, ...,$$

can be considered as a time series. This time series has the covariance function

$$
\begin{aligned}
R^{\hat{m}}(s,t) &= Cov(\hat{m}(s); \hat{m}(t)) \\
&= D[\hat{\beta}_1(X_n)] + Cov(\hat{\beta}_1(X_n); \hat{\beta}_2(X_n))(s+t) \\
&\quad + D[\hat{\beta}_2(X_n)]st
\end{aligned}
$$

and we see that $\hat{m}(.)$ is not covariance stationary. Next it can be seen that

$$Cov(\hat{\beta}_1(X_n); \hat{\beta}_2(X_n)) = -\frac{3}{2n+1}D[\hat{\beta}_1(X_n)] < 0$$

and

$$D[\hat{\beta}_2(X_n)] = \frac{6}{(2n+1)(n+1)}D[\hat{\beta}_1(X_n)].$$

Using these expressions we can write for the variance $D[\hat{m}(.)]$:

$$D[\hat{m}(t)] = D[\hat{\beta}_1(X_n)]\left\{1 - \frac{6[(n+1)t - t^2]}{(2n+1)(n+1)}\right\}.$$

It is easy to show that

$$D[\hat{m}(t)] \leq D[\hat{m}(1)] = D[\hat{m}(n)] = D[\hat{\beta}_1(X_n)]\left(1 - \frac{6n}{(2n+1)(n+1)}\right)$$

and

$$\min_{1 \leq t \leq n} D[\hat{m}(t)] = D[\hat{m}\left(\frac{n+1}{2}\right)] = D[\hat{\beta}_1(X_n)]\left(1 - \frac{3(n+1)^2}{2(2n+1)(n+1)}\right).$$

Thus the smallest variance is in the middle and the largest at the margins of the estimated linear trend $\hat{m}(t); t = 1, 2, ..., n$.

Example 3.2.3. Let $X(.)$ be a time series with a quasiperiodic mean value

$$X(t) = \beta_1 + \sum_{i=1}^{k}(\beta_i^1 \cos \lambda_i^0 t + \beta_i^2 \sin \lambda_i^0 t) + \varepsilon(t); t \in T,$$

where $k \geq 1$, $\beta = (\beta_1, \beta_1^1, ..., \beta_k^1, \beta_1^2, ..., \beta_k^2)' \in E^{2k+1}$ are unknown parameters, and $\lambda_i^0; i = 1, 2, ..., k$, are known frequencies. Then the finite observation X_n of $X(.)$ can be written as

$$X_n = F_n \beta + \varepsilon_n, E[\varepsilon_n] = 0, \beta \in E^{2k+1}, Cov(X_n) = \Sigma_n,$$

where the $n \times (2k+1)$ design matrix F_n is given by

$$F_n' = \begin{pmatrix} 1 & 1 & \cdots & 1 \\ \cos \lambda_1^0 & \cos \lambda_1^0 2 & \cdots & \cos \lambda_1^0 n \\ \sin \lambda_1^0 & \sin \lambda_1^0 2 & \cdots & \sin \lambda_1^0 n \\ . & . & \cdots & \\ \cos \lambda_k^0 & \cos \lambda_k^0 2 & \cdots & \cos \lambda_k^0 n \\ \sin \lambda_k^0 & \sin \lambda_k^0 2 & \cdots & \sin \lambda_k^0 n \end{pmatrix}.$$

If $\lambda_i^0; i = 1, 2, ..., k$, are *Fourier frequencies* which are given by

$$\lambda_i^0 = \frac{2\pi}{n} j_i; i = 1, 2, ..., k,$$

where j_i are some integers, $1 \leq j_i \leq n/2; i = 1, 2, ..., k$ then we get that $F_n' F_n$ is a diagonal matrix

$$F_n' F_n = \begin{pmatrix} n & 0 & \cdots & 0 \\ 0 & n/2 & \cdots & 0 \\ . & . & \cdots & . \\ 0 & 0 & \cdots & n/2 \end{pmatrix}.$$

This follows from the fact that

$$\sum_{t=1}^{n} \cos \lambda_j t = \sum_{t=1}^{n} \sin \lambda_j t$$
$$= \sum_{t=1}^{n} \sin \lambda_j t \cos \lambda_j t = 0$$

and

$$\sum_{t=1}^{n} \cos^2 \lambda_j t = \sum_{t=1}^{n} \sin^2 \lambda_j t = \frac{n}{2}$$

if $\lambda_j; j = 1, 2, ..., k$, are Fourier frequencies. These equalities are consequences of the following statements

$$\sum_{t=1}^{n} (\cos \lambda t + i \sin \lambda t) = \sum_{t=1}^{n} e^{i\lambda t}$$
$$= e^{i\lambda} \frac{1 - e^{in\lambda}}{1 - e^{i\lambda}} = 0$$

if $\lambda = \lambda_j = (2\pi/n)j$, since $e^{in\lambda_j} = e^{i2\pi j} = \cos 2\pi j + i \sin 2\pi j = 1$; $j = 1, 2, ..., n/2$. Next

$$\sum_{t=1}^{n} \sin \lambda_j t \cos \lambda_j t = \frac{1}{2} \sum_{t=1}^{n} \sin 2\lambda_j t = 0$$

and

$$\sum_{t=1}^{n} \cos^2 \lambda_j t = \frac{1}{2} \sum_{t=1}^{n} (1 + \cos 2\lambda_j t) = \frac{n}{2},$$

$$\sum_{t=1}^{n} \sin^2 \lambda_j t = \frac{1}{2} \sum_{t=1}^{n} (1 - \cos 2\lambda_j t) = \frac{n}{2}.$$

Thus we have

$$(F_n' F_n)^{-1} = \begin{pmatrix} 1/n & 0 & ... & 0 \\ 0 & 2/n & ... & 0 \\ . & . & ... & . \\ 0 & 0 & ... & 2/n \end{pmatrix}$$

and the OLSE $\hat{\beta}$ of β is given by

$$\hat{\beta}_1(X_n) = \frac{1}{n} \sum_{t=1}^{n} X(t) = \bar{X}_n,$$

$$\hat{\beta}_i^1(X_n) = \frac{2}{n} \sum_{t=1}^{n} X(t) \cos \lambda_i t,$$

and

$$\hat{\beta}_i^2(X_n) = \frac{2}{n} \sum_{t=1}^{n} X(t) \sin \lambda_i t; 1 = 1, 2, ..., k.$$

It is clear that $(F_n' F_n)^{-1}$; $n = 1, 2, ...$, are of the order $O(1/n)$ and thus the conditions $\lim_{n \to \infty} 1/n (\sum_{s=1}^{n} \sum_{t=1}^{n} R^2(s,t))^{1/2} = 0$ or $\lim_{t \to \infty} R(t) = 0$ are sufficient for consistency of the OLSE $\hat{\beta}$ of parameter β of any quasiperiodic mean value with known Fourier frequencies.

In Fuller (1976) the limiting distribution of standardized OLSE $\hat{\beta}$ is given by the following assumptions on the LRM

$$X(t) = \sum_{i=1}^{k} \beta_i f_i(t) + \varepsilon(t); t \in T, \beta \in E^k.$$

Let us assume that

$$\lim_{n\to\infty} \|f_i\|_n^2 \;=\; \lim_{n\to\infty} \sum_{t=1}^{n} f_i^2(t) = \infty; i = 1, 2, ..., k,$$

$$\lim_{n\to\infty} \frac{f_i(n)}{\|f_i\|_n^2} \;=\; 0; i = 1, 2, ..., k,$$

$$\lim_{n\to\infty} \frac{\sum_{s=1}^{n-t} f_i(s) f_j(s+t)}{\|f_i\|_n \|f_j\|_n} \;=\; a_{t,ij}; t = 0, 1, ..., i, j = 1, 2, ..., k,$$

and let $F'F$ be p.d. for all $n > k$. Let

$$D_n = diag(\|f_1\|_n, ..., \|f_k\|_n)$$

and let A_0, defined by

$$A_0 = \lim_{n\to\infty} D_n^{-1} F' F D_n^{-1},$$

be a nonsingular matrix. Under these assumptions the following theorem is proved in Fuller (1976).

Theorem 3.2.2. *Let $\varepsilon(.)$ be a stationary time series defined by*

$$\varepsilon(t) = \sum_{j=0}^{\infty} b_j Y(t - j); t \in T,$$

where $\{b_j\}_{j=0}^{\infty}$ is absolute summable and the $Y(t)$ are independent random variables with $E[Y(t)] = 0, D[Y(t)] = \sigma^2$, and with distribution functions $F_t(y) = Prob(Y(t) < y); y \in E^1$, such that

$$\lim_{\delta\to\infty} \sup_t \int_{|y|>\delta} y^2 dF_t(y) = 0.$$

Let the $k \times k$ matrix B have elements

$$B_{ij} = \sum_{t=-\infty}^{\infty} a_{t,ij} R^Y(t)$$

and let B be nonsingular. Then the estimator $D_n(\hat{\beta}(X) - \beta)$ is asymptotically normal with mean value zero and with covariance matrix $A_0^{-1} B A_0^{-1}$, and we write

$$D_n(\hat{\beta}(X) - \beta) \sim N_k(0, A_0^{-1} B A_0^{-1}).$$

The OLSE $\hat{\beta}$ in any LRM can be written in the following form

$$\begin{aligned} \hat{\beta}(X) &= (F'F)^{-1}F'X = (F'F)^{-1}F'(F\beta + \varepsilon) \\ &= \hat{\beta}(\varepsilon) = \beta + (F'F)^{-1}F'\varepsilon \end{aligned}$$

and we see that the statistical properties of $\hat{\beta}$ depend on the design matrix F which does not depend on β and on the properties of the vector of errors ε. All properties of $\hat{\beta}$ can be deduced from the last equality, although the random vector ε is unobservable.

This approach can also be used in NRMs

$$X(t) = m_\gamma(t) + \varepsilon(t); t \in T, \gamma \in \Upsilon,$$

in which we can write a finite observation X of $X(.)$ as

$$X = m_\gamma + \varepsilon; \gamma \in \Upsilon.$$

The OLSE $\hat{\gamma}$ of γ, defined by

$$\hat{\gamma}(x) = \arg\min_\gamma \|x - m_\gamma\|^2 = \arg\min_\gamma \sum_{t=1}^n (x(t) - m_\gamma(t))^2; x \in \mathcal{X},$$

we can find as a solution of the normal equations

$$\frac{\partial}{\partial \gamma_i} \sum_{t=1}^n (x(t) - m_\gamma(t))^2 \mid_{\gamma=\hat{\gamma}} = 0; i = 1, 2, ..., k.$$

These equations can be written as

$$\sum_{t=1}^n (x(t) - m_\gamma(t)) \frac{\partial m_\gamma(t)}{\partial \gamma_i} \mid_{\gamma=\hat{\gamma}} = 0; i = 1, 2, ..., k.$$

Denoting by F_{γ_0} the $n \times k$ matrix with components

$$F_{\gamma_0, ti} = \frac{\partial m_\gamma(t)}{\partial \gamma_i} \mid_{\gamma=\gamma_0}; t = 1, 2, .., n, i = 1, 2, ..., k,$$

we can write the normal equations in the matrix form

$$F'_{\hat{\gamma}}(X - m_{\hat{\gamma}}) = 0.$$

As we said in Section 1.3, to compute $\hat{\gamma}(x)$ is a nonlinear problem, and some iterative procedure, such as Gauss-Newton, should be used for this computation. $\hat{\gamma}(x)$ is, in general, a nonlinear function of time series data x for which we have no explicit expression and thus we have no direct tools for deriving the statistical properties of the OLSE $\hat{\gamma}$.

The methodology which can be useful in this context is to derive some approximation $\hat{\gamma}_{ap}$ for $\hat{\gamma}$ and, on the basis of this approximation, to derive the approximated statistical properties of $\hat{\gamma}$. Since the statistical properties of $\hat{\gamma}$ depend, as in the linear case, on properties of the random vector ε, it is sufficient to derive an approximation which depends only on unobservable ε and not on the direct observation X of time series $X(.)$.

A sufficient approximation $\hat{\gamma}_{ap}$ for $\hat{\gamma}$ in terms of ε can be obtained by using the Taylor expansions for m_γ and F_γ around the true value γ_0 of the parameter γ. Using only terms containing the first partial derivatives with respect to $\gamma_i; i = 1, 2, ..., k$, we get, after some algebra, the approximation

$$\hat{\gamma}_{ap}(\varepsilon) = \gamma_0 + (F'F)^{-1}F'\varepsilon + (F'F)^{-1}\left[\begin{pmatrix} \varepsilon'N_1\varepsilon \\ \cdot \\ \cdot \\ \cdot \\ \varepsilon'N_k\varepsilon \end{pmatrix} - \frac{1}{2}F'\begin{pmatrix} \varepsilon'A'H_1A\varepsilon \\ \cdot \\ \cdot \\ \cdot \\ \varepsilon'A'H_nA\varepsilon \end{pmatrix}\right].$$

In this expression we have simplified the notation by writing

$$F_{\gamma_0} = F, A = (F'F)^{-1}F'.$$

The $n \times n$ matrices $N_j; j = 1, 2, ..., k$, are defined by $N_j = \frac{1}{2}(O_j + O_j')$, where

$$O_{j,kl} = \sum_{t=1}^{n}(H_t A)_{jk}M_{tl}; k, l = 1, 2, ..., n.$$

Here $H_t; t = 1, 2, ..., n$, are the $k \times k$ Hessian matrices of m_γ defined by

$$H_{t,ij} = \frac{\partial^2 m_\gamma(t)}{\partial\gamma_i\partial\gamma_j}; i, j = 1, 2, ..., k,$$

and $M = I - F(F'F)^{-1}F'$.

If we compare the expression for the OLSE $\hat{\beta}(\varepsilon)$ with the expression for the approximation $\hat{\gamma}_{ap}(\varepsilon)$ of the OLSE $\hat{\gamma}(\varepsilon)$ we can see that the first part of these expressions is the same, but the expression for the approximation $\hat{\gamma}_{ap}(\varepsilon)$ has an additional term which is equal to the linear combinations of quadratic forms in ε. Another important difference between these two expressions is that the design matrix F in LRM does not depend on the regression parameter β, while all the matrices appearing in $\hat{\gamma}_{ap}$ are functions of the regression parameter γ.

Using the expression for a mean value of a quadratic form we get

$$E_\gamma[\hat{\gamma}_{ap}(\varepsilon)] = \gamma + (F'F)^{-1}\left[\begin{pmatrix} tr(N_1\Sigma) \\ \cdot \\ \cdot \\ \cdot \\ tr(N_k\Sigma) \end{pmatrix} - \frac{1}{2}F'\begin{pmatrix} tr(A'H_1A\Sigma) \\ \cdot \\ \cdot \\ \cdot \\ tr(A'H_nA\Sigma) \end{pmatrix}\right]$$

and we see that the bias of $\hat{\gamma}_{ap}$, which approximates the bias of the OLSE $\hat{\gamma}$, depends on the true value of the regression parameter γ and also on the covariance matrix Σ of X.

In the special case when we assume that $\varepsilon(.)$ is a white noise time series with a variance σ^2 we get $\Sigma = \sigma^2 I_n$ and

$$
E_\gamma[\hat{\gamma}(\varepsilon)] \approx \gamma - \frac{\sigma^2}{2}(F'F)^{-1}F'\begin{pmatrix} tr((F'F)^{-1}H_1) \\ \cdot \\ \cdot \\ \cdot \\ tr((F'F)^{-1}H_n) \end{pmatrix}; \gamma \in \Upsilon,
$$

which is the result derived by Box (1971). This result follows from the fact that $tr(N_j) = tr(O_j) = 0; j = 1, 2, ..., k$, since $AM = 0$.

The question as to whether the bias of the approximation $\hat{\gamma}_{ap}$ goes to zero for n tending to infinity can be solved by the following assumptions for which we write $F = F_n$ to denote the dependence of F on n.

Assumption 1. *The matrices* $(F'_n F_n)^{-1}$, *depending on* γ, *are of the order* $O(1/n)$.

Assumption 2. *The following limits*

$$
\lim_{n\to\infty}\frac{1}{n}\sum_{t=1}^{n}\frac{\partial m_\gamma(t)}{\partial \gamma_i}\frac{\partial^2 m_\gamma(t)}{\partial \gamma_j \partial \gamma_k}
$$

and

$$
\lim_{n\to\infty}\frac{1}{n}\sum_{t=1}^{n}\frac{\partial^2 m_\gamma(t)}{\partial \gamma_i \partial \gamma_j}\frac{\partial^2 m_\gamma(t)}{\partial \gamma_k \partial \gamma_l}
$$

exist and are finite for every fixed i, j, k, l *and for every* $\gamma \in \Upsilon$.

The proof of the following theorem can be found in Štulajter (1992b).

Theorem 3.2.3. *Let Assumptions 1 and 2 hold and let, for the covariance matrix* Σ_n *of the vector* X_n, $\lim_{n\to\infty} 1/n \|\Sigma_n\| = 0$. *Then for the bias of* $\hat{\gamma}$ *we can write*

$$
\lim_{n\to\infty} E_\gamma[\hat{\gamma}(X_n)] - \gamma \approx \lim_{n\to\infty} E_\gamma[\hat{\gamma}_{ap}(\varepsilon_n)] - \gamma = 0.
$$

From this theorem we get that for $\varepsilon(.)$, a white noise time series, or for a covariance stationary time series $\varepsilon(.)$ with an autocovariance function $R(.)$ for which $\lim_{t\to\infty} R(t) = 0$, the OLSE $\hat{\gamma}$ can be, under Assumptions 1 and 2, considered as an asymptotically unbiased estimator of parameter γ.

Using the expression

$$
E[\varepsilon'B\varepsilon\varepsilon'C\varepsilon] = 2tr(B\Sigma C\Sigma) + tr(B\Sigma)tr(C\Sigma)
$$

which holds if $\varepsilon \sim N_n(0, \Sigma)$ we can derive an expression for the $MSE[\hat{\gamma}_{ap}]$

defined by

$$MSE[\hat{\gamma}_{ap}] = E_\gamma[(\hat{\gamma}_{ap} - \gamma)(\hat{\gamma}_{ap} - \gamma)'].$$

The precise expression for a fixed n is of no great practical importance and can be found in Štulajter (1992b). The following theorem, under which the approximation $\hat{\gamma}_{ap}$ is a consistent estimator of γ, is also proven there.

Theorem 3.2.4. *Let Assumptions 1 and 2 hold and let* $\varepsilon_n \sim N_n(0, \Sigma_n)$, *where* $\lim\limits_{n\to\infty} 1/n \|\Sigma_n\| = 0$. *Then*

$$\lim_{n\to\infty} E_\gamma[(\hat{\gamma}_{ap}(\varepsilon_n) - \gamma)(\hat{\gamma}_{ap}(\varepsilon_n) - \gamma)'] = 0.$$

Since $\hat{\gamma}_{ap}$ is an approximation of the OLSE $\hat{\gamma}$, it can be said that this theorem gives sufficient conditions under which $\hat{\gamma}$ is consistent estimator of γ. These conditions for NRMs are similar to those appearing in LRMs, mainly in the case when $\varepsilon(.)$ is a white noise time series with the variance σ^2. Then we have

$$
\begin{aligned}
MSE_\gamma[\hat{\gamma}_{ap}(\varepsilon)] = {} & \sigma^2(F_n'F_n)^{-1} + \sigma^4(F_n'F_n)^{-1}[\sum_{s,t=1}^n M_{st}H_s(F_n'F_n)^{-1}H_t \\
& + \frac{1}{2}\sum_{s,t=1}^n tr(H_s(F_n'F_n)^{-1}H_t(F_n'F_n)^{-1})B_{s,t} \\
& + \frac{1}{4}\sum_{s,t=1}^n tr(H_s(F_n'F_n)^{-1})tr(H_t(F_n'F_n)^{-1})B_{s,t}](F_n'F_n)^{-1},
\end{aligned}
$$

where $B_{s,t}; s,t = 1, 2, ..., n$, are $k \times k$ matrices with the elements

$$(B_{s,t})_{ij} = \frac{\partial m_\gamma(s)}{\partial \gamma_i}\frac{\partial m_\gamma(t)}{\partial \gamma_j}; i, j = 1, 2, ..., k.$$

We see that the first part of this expression is the same as in LRMs and thus the matrices $(F_n'F_n)^{-1}$ again play an important role in the problem of consistency of $\hat{\gamma}_{ap}$.

Example 3.2.4. Let us consider the time series $X(.)$ given by

$$X(t) = m_\gamma(t) + \varepsilon(t); t \in T, \gamma \in \Upsilon,$$

where

$$m_\gamma(t) = \gamma_1 e^{\gamma_2 t}; t \in T, \gamma = (\gamma_1, \gamma_2)' \in \Upsilon.$$

It is easy to show that the 2×2 matrices $F_n' F_n$ are given by

$$F_n' F_n = \begin{pmatrix} \sum\limits_{t=1}^{n} e^{2\gamma_2 t} & \gamma_1 \sum\limits_{t=1}^{n} t e^{2\gamma_2 t} \\ \gamma_1 \sum\limits_{t=1}^{n} t e^{2\gamma_2 t} & \gamma_1 \sum\limits_{t=1}^{n} t^2 e^{2\gamma_2 t} \end{pmatrix}.$$

The limit properties of these matrices depend on the values of parameter γ_2. For $\gamma_2 = 0$ the time series $X(.)$ is stationary in mean value.

For $\gamma_2 > 0$ all the sums generated the elements of $F_n' F_n$ are divergent and the matrices $(F_n' F_n)^{-1}$ are of the order $O(1/n)$ and also Assumption 2 is fulfilled and thus $\lim\limits_{n \to \theta} MSE_\gamma[\hat{\gamma}_{ap}(\varepsilon)] = 0$ if $\lim\limits_{n \to \infty} 1/n \, \|\Sigma_n\| = 0$.

For $\gamma_2 < 0$ the sums are convergent and there exists $\lim\limits_{n \to \theta} (F_n' F_n)^{-1}$ which is a nonzero matrix and thus

$$\lim\limits_{n \to \theta} MSE_\gamma[\hat{\gamma}_{ap}(\varepsilon)] \neq 0.$$

The OLSE $\hat{\gamma}$ can be, in this case, considered as an inconsistent estimator of γ. These results are consequences of the fact that the mean value functions $m_\gamma(t) = \gamma_1 e^{\gamma_2 t}$ and $m_{\gamma'}(t) = \gamma_1' e^{\gamma_2' t}$ are for $\gamma_2 < 0$, and $\gamma_2' < 0$ and for large values of n practically the same, and thus increasing, number of experiments n does not give new information about the unknown parameter γ_2. This is not true if $\gamma_2 > 0$ and $\gamma_2' > 0$. In this case the distance between $m_\gamma(t)$ and $m_{\gamma'}(t)$ increases as t increases and thus every new observation of $X(.)$ gives us important information about the value of γ_2.

To illustrate the situation numerically let us consider the case when $\gamma_1 = 1$ and $\gamma_2 = -0.5$. Then we get

$$(F_{30}' F_{30})^{-1} = \begin{pmatrix} 6.3887 & -2.9523 \\ -2.9523 & 1.8662 \end{pmatrix}$$

and it can be verified that the matrices $(F_n' F_n)^{-1}$ are practically the same for $n \geq 30$.

If $\gamma_1 = 1$ and $\gamma_2 = 0.5$, we get

$$(F_{30}' F_{30})^{-1} = \begin{pmatrix} 5.4619 \times 10^{-11} & -1.8547 \times 10^{-12} \\ -1.8547 \times 10^{-12} & 6.3047 \times 10^{-14} \end{pmatrix}.$$

It can be seen that the convergence of $(F_n' F_n)^{-1}$ to zero is very rapid in this case and for $n = 30$ the matrix $(F_n' F_n)^{-1}$ is very small, practically equal to zero.

Example 3.2.5. Let $X(.)$ be a time series with the logistic trends

$$m_\gamma(t) = \frac{\gamma_1}{1 + \gamma_2 \gamma_3^t}; t \in T, \gamma = (\gamma_1, \gamma_2, \gamma_3)' \in \Gamma = (-\infty, \infty) \times (0, \infty) \times (0, 1).$$

Then the components of the symmetric matrices $F_n' F_n$ are given by

$$(F_n' F_n)_{11} = \sum_{t=1}^{n} 1/(1 + \gamma_2 \gamma_3^t)^2,$$

$$(F_n' F_n)_{12} = -\gamma_1 \sum_{t=1}^{n} \gamma_3^t/(1 + \gamma_2 \gamma_3^t)^3,$$

$$(F_n' F_n)_{13} = -\gamma_1 \gamma_2 \sum_{t=1}^{n} t\gamma_3^t/(1 + \gamma_2 \gamma_3^t)^3,$$

$$(F_n' F_n)_{22} = \gamma_1^2 \sum_{t=1}^{n} \gamma_3^{2t}/(1 + \gamma_2 \gamma_3^t)^4,$$

$$(F_n' F_n)_{23} = \gamma_1^2 \gamma_2 \sum_{t=1}^{n} t\gamma_3^{2t-1}/(1 + \gamma_2 \gamma_3^t)^4,$$

$$(F_n' F_n)_{33} = \gamma_1^2 \gamma_2^2 \sum_{t=1}^{n} t^2 \gamma_3^{2t-2}/(1 + \gamma_2 \gamma_3^t)^4.$$

The properties of $(F_n' F_n)^{-1}$ we shall study numerically for $\gamma_1 = 1, \gamma_2 = 1$, and $\gamma_3 = 0.5$. Then it can be shown numerically that

$$\lim_{n \to \infty} (F_n' F_n)^{-1} \approx \begin{pmatrix} 0 & 0 & 0 \\ 0 & 48.17 & -11.07 \\ 0 & -11.07 & 3.31 \end{pmatrix}$$

and we have that the OLSE $\hat{\gamma}$ is not a consistent estimator. Mainly, the $D_\gamma[\hat{\gamma}_{2ap}]$ is a relatively large number by the given $\gamma = (1, 1, .5)'$.

We get similar results for $\gamma = (1, 1, .8)'$ when

$$\lim_{n \to \infty} (F_n' F_n)^{-1} \approx \begin{pmatrix} 0 & 0 & 0 \\ 0 & 8.59 & -1.05 \\ 0 & -1.05 & .19 \end{pmatrix}.$$

But for $\gamma = (1, 1, .1)'$ we get $\det(F_{30}' F_{30}) = 0$ and only for large n $\det(F_n' F_n) > 0$. It can be shown that for this value of the parameter γ variances for OLSEs of γ_2 and γ_3 are extremely large, since

$$\lim_{n \to \infty} (F_n' F_n)^{-1} \approx \begin{pmatrix} 0 & 0 & 0 \\ 0 & 10565 & -1027.5 \\ 0 & -1027.5 & 101.3 \end{pmatrix}.$$

Example 3.2.6. Let us consider a time series $X(.)$ modeled by

$$X(t) = \gamma_1 + \gamma_2 \cos \lambda t + \varepsilon(t); t \in T,$$

where the unknown parameters $\gamma = (\gamma_1, \gamma_2, \lambda)' \in \Upsilon = (-\infty, \infty)^2 \times \langle -\pi, \pi \rangle$.
Then the matrices $F_n' F_n$ are given by

$$
F_n' F_n = \begin{pmatrix}
n & \sum\limits_{t=1}^{n} \cos \lambda t & -\gamma_2 \sum\limits_{t=1}^{n} t \sin \lambda t \\
\sum\limits_{t=1}^{n} \cos \lambda t & \sum\limits_{t=1}^{n} \cos^2 \lambda t & -\gamma_2 \sum\limits_{t=1}^{n} t \sin \lambda t \cos \lambda t \\
-\gamma_2 \sum\limits_{t=1}^{n} t \sin \lambda t & -\gamma_2 \sum\limits_{t=1}^{n} t \sin \lambda t \cos \lambda t & \gamma_2^2 \sum\limits_{t=1}^{n} t^2 \sin^2 \lambda t
\end{pmatrix}.
$$

In the special case when λ is one of the Fourier frequencies, $\lambda = \lambda_j = (2\pi/n)j$; for some $j = 1, 2, ..., n/2$ we get

$$
F_n' F_n = \begin{pmatrix}
n & 0 & -\gamma_2 \sum\limits_{t=1}^{n} t \sin \lambda t \\
0 & n/2 & -\gamma_2 \sum\limits_{t=1}^{n} t \sin \lambda t \cos \lambda t \\
-\gamma_2 \sum\limits_{t=1}^{n} t \sin \lambda t & -\gamma_2 \sum\limits_{t=1}^{n} t \sin \lambda t \cos \lambda t & \gamma_2^2 \sum\limits_{t=1}^{n} t^2 \sin^2 \lambda t
\end{pmatrix}.
$$

Let us study the properties of $(F_n' F_n)^{-1}$ numerically. Let $\gamma_2 = 1$ and let $\lambda = \pi/5$. Then we get that for $n = 30$:

$$
(F_{30}' F_{30})^{-1} = \begin{pmatrix}
3.38 \times 10^{-2} & -2.40 \times 10^{-4} & 3.49 \times 10^{-4} \\
-2.40 \times 10^{-4} & 6.67 \times 10^{-2} & -1.56 \times 10^{-4} \\
3.49 \times 10^{-4} & -1.56 \times 10^{-4} & 2.27 \times 10^{-4}
\end{pmatrix}
$$

and we see that the variances of the OLSE of γ are very small for a relatively small number $n = 30$ of observations. For $n = 300$ the matrix $(F_{300}' F_{300})^{-1}$ is practically equal to a zero matrix.

Let us consider a case when λ is not a Fourier frequency, let $n = 30$ and let $\lambda = 3\pi/16 = .1875\pi$. Then

$$
(F_{30}' F_{30})^{-1} = \begin{pmatrix}
.03 & 4.29 \times 10^{-3} & -2.32 \times 10^{-4} \\
4.29 \times 10^{-3} & 7.04 \times 10^{-2} & 2.46 \times 10^{-6} \\
-2.32 \times 10^{-4} & 2.46 \times 10^{-6} & 1.96 \times 10^{-4}
\end{pmatrix}
$$

and we have a similar result as for the Fourier frequency $\lambda = 3\pi/15$. Again, for $n = 300$ the matrix $(F_{300}' F_{300})^{-1}$ is practically equal to a zero matrix.

We get similar results for the models

$$
X(t) = \gamma_1 + \gamma_2 \cos \lambda t + \gamma_3 \sin \lambda t + \varepsilon(t); t \in T,
$$

where the unknown parameter $\gamma = (\gamma_1, \gamma_2, \gamma_3, \lambda)' \in \Upsilon = (-\infty, \infty)^3 \times \langle -\pi, \pi \rangle$. For $\gamma_1 = \gamma_2 = \gamma_3 = 1$ and for $\lambda = \pi/5$ we get

$$
(F_{30}' F_{30})^{-1} \approx \begin{pmatrix}
3.53 \times 10^{-2} & 1.63 \times 10^{-2} & -1.53 \times 10^{-2} & 0 \\
1.63 \times 10^{-2} & .20 & -.12 & 0 \\
-1.53 \times 10^{-2} & -.12 & .18 & 0 \\
0 & 0 & 0 & 0
\end{pmatrix}
$$

and for $n = 300$ the matrix $(F_{300}'F_{300})^{-1}$ is again practically equal to a zero matrix.

It can also be expected that in a *quasiperiodic* NRM

$$X(t) = \gamma_1 + \sum_{i=1}^{k} (\beta_i^1 \cos \lambda_i t + \beta_i^2 \sin \lambda_i t) + \varepsilon(t); t \in T,$$

where $\gamma = (\gamma_1, \beta_1^1, ..., \beta_k^1, \beta_1^2, ..., \beta_k^2, \lambda_1, ..., \lambda_k)' = (\gamma_1, \beta^{1\prime}, \beta^{2\prime}, \lambda')' \in \Upsilon \subset E^{3k+1}$ and $k \geq 2$, the matrices $(F_n'F_n)^{-1}$ have the same behavior as for the case $k = 1$.

The important problem in computing the OLSE $\hat{\gamma}$ of γ in any NRM is the problem of choosing an initial value $\hat{\gamma}^{(0)}$ for iterations according to some numerical procedure for computing $\hat{\gamma}(x)$.

For a quasiperiodic NRM this value $\hat{\gamma}^{(0)}$ can be found by using a periodogram. Let

$$X(t) = m(t) + \varepsilon(t); t \in T,$$

be a time series and let $X_n = (X(1), ..., X(n))'$ be a finite observation of $X(.)$. Then the random process $I_n(.)$ defined on $\langle -\pi, \pi \rangle$ by

$$
\begin{aligned}
I_n(\lambda) &= \frac{1}{n} \left[\left(\sum_{t=1}^{n} X(t) \cos \lambda t \right)^2 + \left(\sum_{t=1}^{n} X(t) \sin \lambda t \right)^2 \right] \\
&= \frac{1}{n} \left[\sum_{t=1}^{n} \sum_{s=1}^{n} X(t)X(s)(\cos \lambda t \cos \lambda s + \sin \lambda t \sin \lambda s) \right] \\
&= \frac{1}{n} \left[\sum_{t=1}^{n} \sum_{s=1}^{n} X(t)X(s) \cos \lambda(s - t) \cos \lambda s \right]; \lambda \in \langle -\pi, \pi \rangle,
\end{aligned}
$$

or by

$$I_n(\lambda) = \frac{1}{n} \left| \sum_{t=1}^{n} X(t)e^{i\lambda t} \right|^2 ; \lambda \in \langle -\pi, \pi \rangle,$$

is called *the periodogram*. It is usual to compute the values of a periodogram only at the Fourier frequencies $\lambda_j = (2\pi/n)j; j = 1, 2, ..., n/2$.

The periodogram $I_n(.)$ can be written in the following form

$$I_n(\lambda) = \frac{1}{n} \left\{ \left(\sum_{t=1}^{n} [m(t) + \varepsilon(t)] \cos \lambda t \right)^2 + \left(\sum_{t=1}^{n} [m(t) + \varepsilon(t)] \sin \lambda t \right)^2 \right\},$$

or in the form

$$I_n(\lambda) = \frac{1}{n}\left(\sum_{t=1}^{n} m(t)\cos \lambda t + \sum_{t=1}^{n} \varepsilon(t)\cos \lambda t\right)^2$$

$$+\frac{1}{n}\left(\sum_{t=1}^{n} m(t)\sin \lambda t + \sum_{t=1}^{n} \varepsilon(t)\sin \lambda t\right)^2.$$

Let us consider in more detail the case when the mean value $m(.)$ of $X(.)$ follows a quasiperiodic NRM

$$m(t) = m_\gamma(t) = \gamma_1 + \sum_{i=1}^{k}(\beta_i^1 \cos \lambda_i t + \beta_i^2 \sin \lambda_i t); t = 1, 2, ..., n,$$

where we shall assume first that $k < n/2$ and $\lambda_i; i = 1, 2, ..., k$, are Fourier frequencies which are unknown.

Let λ_j be a Fourier frequency, $\lambda_j \neq \lambda_i$ for all $i = 1, 2, ..., k$. Then we get

$$\sum_{t=1}^{n} m(t)\cos \lambda_j t = \sum_{t=1}^{n} m(t)\sin \lambda_j t = 0 \text{ for all } n.$$

This follows from the derived equalities

$$\sum_{t=1}^{n} \sin \lambda_i t = \sum_{t=1}^{n} \cos \lambda_j t = 0$$

and from the equalities

$$\sum_{t=1}^{n} \cos \lambda_i t \cos \lambda_j t = 0,$$

$$\sum_{t=1}^{n} \sin \lambda_i t \sin \lambda_j t = 0,$$

$$\sum_{t=1}^{n} \sin \lambda_i t \cos \lambda_j t = 0,$$

if λ_j and λ_i are different Fourier frequencies. To show that the last equalities hold we use the equality

$$\sum_{t=1}^{n} e^{it(\lambda-\kappa)} = e^{i(\lambda-\kappa)}\frac{e^{in(\lambda-\kappa)} - 1}{e^{i(\lambda-\kappa)} - 1} = 0,$$

which holds if $(\lambda - \kappa) = (2\pi/n)k; k = ..., -2, -1, 1, 2, ...$ which is fulfilled if $\lambda = \lambda_i$ and $\kappa = \lambda_j$ and λ_i and λ_j are different Fourier frequencies. But

$$e^{it(\lambda - \kappa)} = (\cos \lambda t + i \sin \lambda t)(\cos \kappa t - i \sin \kappa t)$$

and thus we have

$$\sum_{t=1}^{n} \cos \lambda t \cos \kappa t + \sum_{t=1}^{n} \sin \lambda t \sin \kappa t = 0,$$

$$\sum_{t=1}^{n} \sin \lambda t \cos \kappa t - \sum_{t=1}^{n} \cos \lambda t \sin \kappa t = 0.$$

This can be written as

$$\sum_{t=1}^{n} \cos \lambda t \cos \kappa t = -\sum_{t=1}^{n} \sin \lambda t \sin \kappa t,$$

$$\sum_{t=1}^{n} \sin \lambda t \cos \kappa t = \sum_{t=1}^{n} \cos \lambda t \sin \kappa t.$$

Setting $\kappa = -\kappa$ in the first equality and $\lambda = -\lambda$ in the second, we get that

$$\sum_{t=1}^{n} \cos \lambda t \cos \kappa t = \sum_{t=1}^{n} \sin \lambda t \sin \kappa t,$$

$$\sum_{t=1}^{n} \sin \lambda t \cos \kappa t = -\sum_{t=1}^{n} \cos \lambda t \sin \kappa t,$$

from which the equalities, which should be proved, follow.

Thus we have shown that for any Fourier frequency λ such that $\lambda \neq \lambda_i$ for all $i = 1, 2, ..., k$, we can write

$$I_n(\lambda) = \frac{1}{n} \left[\left(\sum_{t=1}^{n} \varepsilon(t) \cos \lambda t \right)^2 + \left(\sum_{t=1}^{n} \varepsilon(t) \sin \lambda t \right)^2 \right] = I_n^{\varepsilon}(\lambda),$$

where $I_n^{\varepsilon}(.)$ is the periodogram of the time series $\varepsilon(.)$.

For the periodogram $I_n^{\varepsilon}(.)$ we can write

$$n I_n^{\varepsilon}(\lambda) = A^2(\lambda) + B^2(\lambda),$$

where

$$A(\lambda) = \sum_{t=1}^{n} \varepsilon(t) \cos \lambda t$$

and

$$B(\lambda) = \sum_{t=1}^{n} \varepsilon(t) \sin \lambda t.$$

Clearly, when $\varepsilon(.)$ is a *Gaussian white noise* $A(\lambda)$ and $B(\lambda)$ are zero mean, normally distributed random variables. Furthermore, for Fourier frequencies λ, we get

$$D[A(\lambda)] = \sigma^2 \sum_{t=1}^{n} \cos^2 \lambda t = \frac{n}{2}\sigma^2$$

and

$$D[B(\lambda)] = \sigma^2 \sum_{t=1}^{n} \sin^2 \lambda t = \frac{n}{2}\sigma^2.$$

Also

$$Cov(A(\lambda); B(\lambda)) = \sigma^2 \sum_{t=1}^{n} \cos \lambda t \sin \lambda t = 0.$$

In this case $\sqrt{2/(n\sigma^2)}A(\lambda)$ and $\sqrt{2/(n\sigma^2)}B(\lambda)$ are, for Fourier frequencies λ, independent standard normal random variables and

$$\frac{2}{n\sigma^2}(A^2(\lambda) + B^2(\lambda)) = \frac{2}{\sigma^2} I_n^\varepsilon(\lambda) \sim \chi_2^2 \text{ distribution.}$$

Thus, if $\varepsilon(.)$ is a Gaussian white noise with variance σ^2, then

$$E[I_n(\lambda_j)] = E[I_n^\varepsilon(\lambda_j)] = \sigma^2$$

and

$$D[I_n(\lambda_j)] = D[I_n^\varepsilon(\lambda_j)] = \sigma^4$$

for all such Fourier frequencies λ_j which are not equal to the Fourier frequencies $\lambda_i; i = 1, 2, ..., k$.

Now let $\lambda_j = \lambda_i$ for some $i = 1, 2, ..., k$. Then we have, using the derived equalities,

$$I_n(\lambda_j) = \frac{1}{n}\left[\sum_{t=1}^{n}\left(\gamma_1 + \sum_{i=1}^{k}(\beta_i^1 \cos \lambda_i t + \beta_i^2 \sin \lambda_i t) + \varepsilon(t)\right)\cos \lambda_j t\right]^2$$

$$+ \frac{1}{n}\left[\sum_{t=1}^{n}\left(\gamma_1 + \sum_{i=1}^{k}(\beta_i^1 \cos \lambda_i t + \beta_i^2 \sin \lambda_i t) + \varepsilon(t)\right)\sin \lambda_j t\right]^2$$

$$= \quad I_n(\lambda_i) = \frac{1}{n} \left(\beta_i^1 \sum_{t=1}^{n} \cos^2 \lambda_i t + \sum_{t=1}^{n} \varepsilon(t) \cos \lambda_i t \right)^2$$

$$+ \frac{1}{n} \left(\beta_i^1 \sum_{t=1}^{n} \sin^2 \lambda_i t + \sum_{t=1}^{n} \varepsilon(t) \sin \lambda_i t \right)^2$$

$$= \quad \frac{1}{n} \left[\left(\frac{n}{2} \beta_i^1 + \sum_{t=1}^{n} \varepsilon(t) \cos \lambda_i t \right)^2 + \left(\frac{n}{2} \beta_i^2 + \sum_{t=1}^{n} \varepsilon(t) \sin \lambda_i t \right)^2 \right]$$

$$= \quad n \left[\left(\frac{1}{2} \beta_i^1 + \frac{1}{n} \sum_{t=1}^{n} \varepsilon(t) \cos \lambda_i t \right)^2 + \left(\frac{1}{2} \beta_i^2 + \frac{1}{n} \sum_{t=1}^{n} \varepsilon(t) \sin \lambda_i t \right)^2 \right].$$

Since

$$E \left[\frac{1}{n} \sum_{t=1}^{n} \varepsilon(t) \cos \lambda_i t \right] = E \left[\frac{1}{n} \sum_{t=1}^{n} \varepsilon(t) \sin \lambda_i t \right] = 0,$$

$$D \left[\frac{1}{n} \sum_{t=1}^{n} \varepsilon(t) \cos \lambda_i t \right] = D \left[\frac{1}{n} \sum_{t=1}^{n} \varepsilon(t) \sin \lambda_i t \right] = \frac{\sigma^2}{2n},$$

it follows that

$$\lim_{n \to \infty} I_n(\lambda_i) = +\infty \text{ for all } i = 1, 2, ..., k.$$

Thus the periodogram gives, in the case when $\varepsilon(.)$ is a white noise, the approximations $\hat{k}$ and $\hat{\lambda}_i^{(0)}; i = 1, 2, ..., \hat{k}$, for the number k of frequencies and also for values $\lambda_i; i = 1, 2, ..., \hat{k}$, of frequencies which should be included into the quasiperiodic regression model. The values $\hat{\lambda}_i^{(0)}; i = 1, 2, ..., \hat{k}$, are the ordinates, Fourier frequencies, of the periodogram $I_n(.)$ at which it takes its local maxima and $\hat{k}$ is the number of these local maxima. There exists the Fisher test for testing a hypothesis as to whether the given local maximum takes a significant value and whether the corresponding Fourier frequency should be included into the regression model. This can be found in Anděl (1976).

The quasiperiodic regression model with estimated values $\hat{k}$ and $\hat{\lambda}_i^{(0)}$, where $i = 1, 2, ..., \hat{k}$ is an LRM with unknown parameters $\gamma_1, \beta_i^1, \beta_i^2; i = 1, 2, ..., \hat{k}$, which should be estimated. The OLSEs $\hat{\gamma}_1, \hat{\beta}_i^1, \hat{\beta}_i^2$ of $\gamma_1, \beta_i^1, \beta_i^2; i = 1, 2, ..., \hat{k}$, can be used as initial estimators and together with $\hat{k}$ and $\hat{\lambda}_i^{(0)}; i = 1, 2, ..., \hat{k}$, we have an estimator $\hat{\gamma}^{(0)}$ of γ which can be used initially in some iterative procedure for computing the OLSE $\hat{\gamma}$ of γ.

Example 3.2.7. Let us consider the NRM

$$X(t) = \gamma_1 + \gamma_2 t + \sum_{i=1}^{2} (\beta_i^1 \cos \lambda_i t + \beta_i^2 \sin \lambda_i t) + \varepsilon(t); t \in T,$$

where $\gamma = (\gamma_1, \gamma_2, \beta_1^1, \beta_1^2, \beta_2^1, \beta_2^2, \lambda_1, \lambda_2)'$ is an unknown vector of regression parameters and $\varepsilon(.)$ is an AR(1) time series given by

$$\varepsilon(t) = \rho \varepsilon(t-1) + e(t)$$

with a white noise $e(.)$ having variance $\sigma^2 = 1$.

Simulated data following this NRM were considered with different values of an autoregression parameter ρ and a given value of γ. For every fixed value of parameters ρ and γ one observation of the length $n = 51$, one of the length $n = 101$, and one of the length $n = 149$ were simulated. The modified Marquard method was used to compute the OLSE $\hat{\gamma}$ of γ.

A comparison of the OLSE $\hat{\gamma}$ and the approximate OLSE $\hat{\gamma}_{ap}$ was done by Štulajter and Hudáková (1991a). It was shown that $\hat{\gamma}$ and $\hat{\gamma}_{ap}$ are nearly the same in many cases. The aim of the simulation study is to investigate the influence of different values of the parameter ρ on the OLSE $\hat{\gamma}$ and the dependence of this influence on n, the length of an observation.

The initial values for iterations were found as follows. First, from X the OLSEs $\hat{\gamma}_1^{(0)}$ and $\hat{\gamma}_2^{(0)}$ were found. Then the periodogram, based on partial residuals $X(t) - \hat{\gamma}_1^{(0)} + \hat{\gamma}_2^{(0)} t; t = 1, 2, ..., n$, was computed and the frequencies $\hat{\lambda}_i^{(0)}; i = 1, 2$, in which there are the two greatest values of the periodogram, were found. In the model

$$X(t) - \hat{\gamma}_1^{(0)} + \hat{\gamma}_2^{(0)} t = \sum_{i=1}^{2} (\beta_i^1 \cos \hat{\lambda}_i^{(0)} t + \beta_i^2 \sin \hat{\lambda}_i^{(0)} t) + \varepsilon(t)$$

the ordinary least squares method was used for finding $\hat{\beta}_i^{1(0)}$ and $\hat{\beta}_i^{2(0)}; i = 1, 2$.

The value $\hat{\gamma}^{(0)} = (\hat{\gamma}_1^{(0)}, \hat{\gamma}_2^{(0)}, \hat{\beta}_1^{1(0)}, \hat{\beta}_1^{2(0)}, \hat{\beta}_2^{1(0)}, \hat{\beta}_2^{2(0)}, \hat{\lambda}_1^{(0)}, \hat{\lambda}_2^{(0)})'$ of an unknown parameter γ was used as an initial value for computing the OLSE $\hat{\gamma}$ of γ using the Marquard method. The OLSEs, each computed from one simulation of the corresponding length n, are given in the following tables.

Table 3.1. Estimates of γ for $n = 51$.

ρ	-0.99	-0.60	-0.20	0	0.2	0.60	0.99
$\gamma_1 = 3.00$	3.02	3.09	2.83	2.98	3.09	3.23	0.71
$\gamma_2 = 2.00$	2.01	1.99	2.00	2.00	2.00	1.99	2.08
$\lambda_1 = 0.75$	0.99	0.75	0.75	0.75	0.75	0.74	0.74
$\lambda_2 = 0.25$	0.75	0.24	0.25	0.25	0.24	0.24	0.25
$\beta_1^1 = 4.00$	-1.67	3.21	4.19	4.32	3.91	3.92	3.78
$\beta_1^2 = 3.00$	-10.59	3.15	2.40	2.78	3.02	2.50	3.20
$\beta_2^1 = 2.00$	3.60	2.11	1.52	2.31	1.88	2.43	0.56
$\beta_2^2 = 4.00$	2.77	4.01	3.74	3.95	3.66	3.93	4.29

Table 3.2. Estimates of γ for $n = 101$.

ρ	-0.99	-0.60	-0.20	0	0.20	0.60	0.99
$\gamma_1 = 3.00$	3.20	2.92	2.94	2.89	3.18	3.47	3.78
$\gamma_2 = 2.00$	1.99	2.00	2.00	2.00	1.99	1.99	1.98
$\lambda_1 = 0.75$	0.99	0.75	0.75	0.75	0.74	0.75	0.75
$\lambda_2 = 0.25$	0.75	0.24	0.25	0.25	0.24	0.25	0.25
$\beta_1^1 = 4.00$	-0.70	3.60	3.88	3.73	4.13	3.75	3.75
$\beta_1^2 = 3.00$	-10.51	3.52	3.72	3.29	2.88	3.24	3.30
$\beta_2^1 = 2.00$	3.66	2.15	2.05	2.35	2.43	2.19	1.51
$\beta_2^2 = 4.00$	2.90	3.86	3.86	3.74	3.61	4.01	4.40

Table 3.3. Estimates of γ for $n = 149$.

ρ	-0.99	-0.60	-0.20	0	0.20	0.60	0.99
$\gamma_1 = 3.00$	3.04	2.97	2.86	3.09	2.95	3.10	3.38
$\gamma_2 = 2.00$	1.99	2.00	2.00	1.99	1.99	1.99	1.99
$\lambda_1 = 0.75$	0.74	0.75	0.75	0.75	0.75	0.75	0.75
$\lambda_2 = 0.25$	0.25	0.25	0.25	0.25	0.25	0.25	0.25
$\beta_1^1 = 4.00$	4.22	4.14	3.92	3.85	4.18	4.07	3.97
$\beta_1^2 = 3.00$	2.68	2.93	3.45	2.80	3.01	3.01	2.97
$\beta_2^1 = 2.00$	2.05	1.77	1.82	2.15	1.95	2.23	1.95
$\beta_2^2 = 4.00$	4.01	3.98	4.09	3.85	3.83	3.87	4.27

We can see from the tables that the only difficulty with estimation is for $\rho = -0.99$, where the influence of the spectral density of an AR(1) process on the periodogram occurs. Here $\lambda_1 = 0.75$ is discovered as a second peak of the periodogram and the estimates of the corresponding βs are 3.60 and 2.77 for $n = 51$ and 3.66 and 2.90 for $n = 101$ instead of 4 and 3 respectively. The value $\hat{\lambda}_1 = 0.99$ is due to the spectral density of the AR(1) time series and also the estimates of the βs correspond to this frequency. This effect does not occur for $n = 149$ or for other values of ρ. The OLSE $\hat{\gamma}$ of γ is also satisfactory for other values of ρ, as we can see from the tables, even for $n = 51$, a relatively small number of observations.

In NRMs

$$X(t) = m_\gamma(t) + \varepsilon(t); t \in T, \gamma \in \Upsilon,$$

in which we can write a finite observation X of $X(.)$ as

$$X = m_\gamma + \varepsilon; \gamma \in \Upsilon, Cov(X) = \Sigma,$$

where Σ is assumed to be known, we can define the WELSE $\hat{\gamma}_\Sigma$ by

$$\hat{\gamma}_\Sigma(x) = \arg\min_\gamma \|x - m_\gamma\|_{\Sigma^{-1}}^2 \, ; x \in \mathcal{X}.$$

These estimators appear by computing the MLE in NRMs with covariance functions following some linear or nonlinear model.

Conditions under which the OLSE $\hat{\gamma}$ and the WELSE γ^* are strongly consistent and asymptotically normal can be found in Gumpertz and Pantulla (1992).

3.3   Estimation of a Covariance Function

In Section 2.2 we described basic models for the covariance functions $R(.,.)$ of time series. Covariance functions can be stationary or may be given by some linear or nonlinear regression model. Examples of such models were also given in Section 2.2. As we have said in the Introduction to this chapter, we also need some model for the mean value of an observed time series for estimating its covariance function. In some cases the mean value of the studied time series is known and is equal to some known function $m(.)$. Using an additive model we can write

$$X(t) = m(t) + \varepsilon(t); E[\varepsilon(t)] = 0, Cov[X(s); X(t)] = R(s, t); s, t \in T,$$

or, for the observation of the length n, we can write

$$X = m + \varepsilon; E[\varepsilon] = 0, Cov(\varepsilon) = \Sigma,$$

where $\Sigma_{st} = R(s, t); s, t = 1, 2, ..., n$.

Let us now study the problem of estimation of a covariance function $R(.,.)$ under the assumption that $X(.)$ is covariance stationary, that is the equality $R(s, t) = R(|s - t|)$ holds for any $s, t \in T$. Then the $n \times n$ covariance matrix $\Sigma = Cov(X)$ has a special form and is given by

$$\Sigma = \begin{pmatrix} R(0) & R(1) & ... & R(n-1) \\ R(1) & R(0) & ... & R(n-2) \\ . & . & ... & . \\ R(n-1) & R(n-2) & ... & R(0) \end{pmatrix}.$$

This matrix can also be written in the form

$$\Sigma = \sum_{t=0}^{n-1} R(t)V_t,$$

where $V_0 = I_n$ and the $n \times n$ matrices $V_t; t = 1, 2, ..., n - 1$, are given by $V_{t,ij} = 1$ if $|i - j| = t$ and $V_{t,ij} = 0$ if $|i - j| \neq t$ for $i, j = 1, 2, ..., n$. These matrices can also be written as

$$V_t = \begin{pmatrix} 0 & I_{n-t} \\ 0 & 0 \end{pmatrix} + \begin{pmatrix} 0 & 0 \\ I_{n-t} & 0 \end{pmatrix}; t = 1, 2, ..., n - 1,$$

with I_{n-t} equal to the $(n - t) \times (n - t)$ identity matrix.

Thus the covariance matrices Σ of X, if X is covariance stationary, are defined by the LRM with matrices $V_t; t = 0, 1, ..., n - 1$, which are known and with an unknown parameter $R = (R(0), R(1), ..., R(n - 1))'$. We can write

$$\Sigma \in \Xi = \left\{ \Sigma_R : \Sigma_R = \sum_{t=0}^{n-1} R(t)V_t; R \in \Upsilon \right\}.$$

The vector R should be estimated on the base of X. This LRM is identical with those introduced in Example 1.4.3.

For estimating the unknown vector R of variance-covariance components we shall use the same principle as was used in Section 1.4 for deriving the DOOLSE. Since we assume that the mean value vector m is known, we use as an initial estimator of Σ the matrix

$$S(X) = (X - m)(X - m)'$$

and define the estimator $\tilde{R} = (\tilde{R}(0), \tilde{R}(1), ..., \tilde{R}(n - 1))'$ of R by

$$\tilde{R}(X) = \arg\min_R \left\| S(X) - \sum_{t=0}^{n-1} R(t)V_t \right\|^2.$$

The components $\tilde{R}(t); t = 0, 1, ..., n - 1$, of $\tilde{R}$ are given, as in Example 1.4.3, by

$$\tilde{R}(t) = \frac{1}{\|V_t\|^2} (S(X), V_t),$$

where we have used the short notation $\tilde{R}(t)$ instead of $\tilde{R}(t)(X)$. These estimators can also be written as

$$\tilde{R}(t) = \frac{1}{n - t} \sum_{s=1}^{n-t} (X(s + t) - m(s + t))(X(s) - m(s)); t = 0, 1, ..., n - 1.$$

Let $B_0 = V_0 = I_n$ and let $B_t = \frac{1}{2}V_t; t = 1, 2, ..., n-1$. Then we get

$$\tilde{R}(t) = \frac{1}{n-t}(S(X), B_t) = \frac{1}{n-t}(X - m)'B_t(X - m)$$

and we can see that the estimators $\tilde{R}(t); t = 0, 1, ..., n-1$, can be considered as quadratic forms in the random vector $X - m$. Using these expressions we can write

$$E_R[\tilde{R}(t)] = \frac{1}{n-t}tr(B_t\Sigma) = R(t); t = 0, 1, ..., n-1,$$

and we see that the estimators $\tilde{R}(t)$ are unbiased for $R(t); t = 0, 1, ..., n-1$.

Variances of these estimators we shall compute under the assumption that $X(.)$ is a Gaussian time series. Then we have

$$D_R[\tilde{R}(t)] = \frac{1}{(n-t)^2}2tr(B_t\Sigma B_t\Sigma).$$

For $t = 0, B_0 = I_n$, and we get

$$
\begin{aligned}
D_R[\tilde{R}(0)] &= \frac{2}{n^2}tr(\Sigma^2) \\
&= \frac{2}{n^2}\|\Sigma\|^2 \\
&= \frac{2}{n^2}\left[nR^2(0) + 2\sum_{t=1}^{n-1}(n-t)R^2(t)\right] \\
&= 2\left[\frac{R^2(0)}{n} + \frac{2}{n}\sum_{t=1}^{n}\left(1 - \frac{t}{n}\right)R^2(t)\right].
\end{aligned}
$$

For $t \neq 0$ we can write $B_t = \frac{1}{2}(K_t + K_t')$, where

$$K_t = \begin{pmatrix} 0 & 0 \\ I_{n-t} & 0 \end{pmatrix}$$

and we get

$$
\begin{aligned}
tr((B_t\Sigma)^2) &= \frac{1}{4}tr(((K_t + K_t')\Sigma)^2) \\
&= \frac{1}{4}[tr((K_t\Sigma)^2) + tr((K_t\Sigma K_t'\Sigma)) + tr(K_t'\Sigma K_t\Sigma) + tr((K_t'\Sigma)^2)] \\
&= \frac{1}{2}[tr((K_t\Sigma)^2) + tr((K_t\Sigma K_t'\Sigma))].
\end{aligned}
$$

Let us write Σ in block form

$$\Sigma = \left(\begin{array}{cc} \Sigma_{1,1} & \Sigma_{1,2} \\ \Sigma_{2,1} & \Sigma_{2,2} \end{array} \right),$$

where

$$\Sigma_{1,2} = \left(\begin{array}{cccc} R(t) & R(t+1) & \ldots & R(n-1) \\ R(t-1) & R(t) & \ldots & R(n-2) \\ . & . & \ldots & . \\ R(n-2t-1) & R(n-2t) & \ldots & R(t) \end{array} \right)$$

is the $(n-t) \times (n-t)$ matrix. Then we get

$$K_t\Sigma = \left(\begin{array}{cc} 0 & 0 \\ \Sigma_{1,1} & \Sigma_{1,2} \end{array} \right)$$

and

$$tr((K_t\Sigma)^2) = tr((\Sigma_{1,2})^2) = (n-t)R^2(t) + 2 \sum_{s=1}^{n-t-1} (n-t-s)R(t+s)R(t-s).$$

We can also write $K_t\Sigma$ as

$$K_t\Sigma = \left(\begin{array}{cc} 0 & 0 \\ C_t & C_{t,2} \end{array} \right),$$

where C_t is the $(n-t) \times (n-t)$ covariance matrix of $(X(1), ..., X(n-t))'$ given by

$$C_t = \left(\begin{array}{cccc} R(0) & R(1) & \ldots & R(n-t-1) \\ R(1) & R(0) & \ldots & R(n-t-2) \\ . & . & \ldots & . \\ R(n-t-1) & R(n-t-2) & \ldots & R(0) \end{array} \right).$$

Writing Σ in another block form,

$$\Sigma = \left(\begin{array}{cc} D_{t,1} & D_{t,2} \\ D_{t,3} & C_t \end{array} \right),$$

we get

$$K_t'\Sigma = \left(\begin{array}{cc} D_{t,3} & C_t \\ 0 & 0 \end{array} \right)$$

and

$$tr((K_t\Sigma K_t'\Sigma)) = tr(C_t^2) = (n-t)R^2(0) + 2(n-t) \sum_{s=1}^{n-t-1} (n-t-s)R^2(s).$$

The derived results can be used in the statement of the following theorem.

Theorem 3.3.1. *Let* $X(.)$ *be a covariance stationary time series with a known mean value function* $m(.)$ *and with an unknown covariance function* $R(.)$. *Then the estimators*

$$\tilde{R}_n(t) = \frac{1}{n-t} \sum_{s=1}^{n-t} (X(s+t) - m(s+t))(X(s) - m(s))$$

are unbiased for $R(t)$ *for every* $t = 0, 1, ..., n-1$. *Moreover, if* $X(.)$ *is Gaussian, then*

$$D_R[\tilde{R}_n(t)] = \frac{1}{n-t} \left(R^2(0) + R^2(t) \right)$$

$$+ \frac{2}{n-1} \sum_{s=1}^{n-t} \left(1 - \frac{s}{n-t} \right) (R^2(s) + R(t+s)R(t-s))$$

and, if $\lim_{t \to \infty} R(t) = 0$, *then* $\lim_{n \to \infty} D_R[\tilde{R}_n(t)] = 0$ *for every* t.

Proof. It is enough to prove only the last statement of the theorem. This follows from the inequality

$$\sum_{s=1}^{n-t} \left(1 - \frac{s}{n-t} \right) R^2(s) \le \sum_{s=1}^{n-t} R^2(s)$$

and from the fact that

$$\lim_{n \to \infty} \frac{1}{n-t} \sum_{s=1}^{n-t} R^2(s) = 0$$

if $\lim_{t \to \infty} R(t) = \lim_{t \to \infty} R^2(t) = 0$. Next, using the Schwarz inequality, we can write

$$\left[\frac{1}{n-t} \sum_{s=1}^{n-t} \left(1 - \frac{s}{n-t} \right) R(t+s)R(t-s) \right]^2$$

$$\le \frac{1}{n-t} \sum_{s=1}^{n-t} \left(1 - \frac{s}{n-t} \right) R^2(t+s) \frac{1}{n-t} \sum_{s=1}^{n-t} \left(1 - \frac{s}{n-t} \right) R^2(t-s)$$

$$\le \frac{1}{n-t} \sum_{s=1}^{n-t} R^2(t+s) \frac{1}{n-t} \sum_{s=1}^{n-t} R^2(t-s).$$

The limit for $n \to \infty$ of the last expression is zero, since for every fixed t we have $\lim_{s \to \infty} R^2(s+t) = \lim_{s \to \infty} R^2(s-t) = 0$ and the theorem is proved.

According to the statement of this theorem the estimators $\tilde{R}_n(t)$ are consistent for every fixed t under a weak condition $\lim_{t \to \infty} R(t) = 0$ on the estimated covariance function. The last condition cannot be verified on the base of the finite observation X, but it is a natural one and it can be assumed to hold in many practical applications of time series models.

The large sample properties of the estimated values of a covariance function $R(.)$ of a time series $X(.)$ having mean value zero are also investigated in Fuller (1976).

It is proved there that in the case when $X(.)$ is a mean value zero stationary time series defined by

$$X(t) = \sum_{j=0}^{\infty} b_j Y(t-j); t \in T,$$

where $\{b_j\}_{j=0}^{\infty}$ is absolutely summable and the $Y(t)$ are independent random variables with $E[Y(t)] = 0, D[Y(t)] = \sigma^2, E[Y(t)^4] = k\sigma^4$ for some constant k and with a finite sixth moment $E[Y(t)^6]$, then the limiting distribution of the $l \times 1$ random vector

$$n^{\frac{1}{2}} \left(\bar{R}(0) - R(0), \bar{R}(1) - R(1), ..., \bar{R}(l) - R(l) \right)$$

is, for any fixed l, multivariate normal with mean zero and covariance matrix V. The estimators $\bar{R}(t)$ are defined by

$$\bar{R}(t) = \frac{1}{n} \sum_{s=1}^{n-t} (X(s+t) - \bar{X})(X(s) - \bar{X}); t = 0, 1, ...,$$

where

$$\bar{X} = \frac{1}{n} \sum_{t=1}^{n} X(t)$$

and for the elements V_{st} of V the following equality

$$\lim_{n \to \infty} (n-t)V_{st} = (k-3)R(s)R(t) + \sum_{u=-\infty}^{\infty} [R(u)R(u-s+t) + R(u+t)R(u-s)]$$

holds for any fixed s and $t, s \geq t \geq 0$.

There are only a few cases in which we can assume that the mean value of the observed time series is known. Let us assume that the mean value $m(.)$ of $X(.)$ is not known and let us assume that it is given by a LRM. Then we shall write

$$X(t) = m_\beta(t) + \varepsilon(t) = \sum_{i=1}^{k} \beta_i f_i(t) + \varepsilon(t); t \in T, \beta \in E^k.$$

Again assume that $X(.)$ is covariance stationary, that is,

$$R(s,t) = R(|s-t|); s,t \in T.$$

Under these assumptions we get that a finite observation X of the length n of $X(.)$ is given by a mixed LRM

$$X = F\beta + \varepsilon; E[\varepsilon] = 0, \beta \in E^k, Cov(X) = \Sigma = \sum_{t=0}^{n-1} R(t)V_t, R \in \Upsilon,$$

where, as before, $V_0 = I_n$ and

$$V_t = \begin{pmatrix} 0 & I_{n-t} \\ 0 & 0 \end{pmatrix} + \begin{pmatrix} 0 & 0 \\ I_{n-t} & 0 \end{pmatrix}; t = 1, 2, ..., n-1.$$

This model was introduced in Example 1.4.3 where the DOOLSE was also derived for the unknown vector $R = (R(0), R(1), .., R(n-1))'$ of values of a covariance function $R(.)$. The DOOLSE $\hat{R} = (\hat{R}(0), \hat{R}(1), ..., \hat{R}(n-1))'$ is defined as follows. Let $\hat{\beta}$ be the OLSE of β, $\hat{\beta}(X) = (F'F)^{-1}F'X$, and let

$$S(X) = (X - F\hat{\beta})(X - F\hat{\beta})'.$$

Then $\hat{R}$ is defined by

$$\hat{R}(X) = \arg\min_{R} \left\| S(X) - \sum_{t=0}^{n-1} R(t)V_t \right\|^2.$$

The components $\hat{R}(t); t = 0, 1, ..., n-1$, of $\hat{R}$ are given by

$$\hat{R}(t) = \frac{1}{\|V_t\|^2}(S(X), V_t),$$

where we have again used the short notation $\hat{R}(t)$ instead of $\hat{R}(t)(X)$. These estimators can also be written as

$$\hat{R}(t) = \frac{1}{n-t}\sum_{s=1}^{n-t}(X(s+t) - \hat{m}(s+t))(X(s) - \hat{m}(s)); t = 0, 1, ..., n-1,$$

where $\hat{m}(t) = m_{\hat{\beta}}(t) = (F\hat{\beta})(t); t = 1, 2, ..., n$, are components of the estimated mean $\hat{m}$ of X.

Let $P = F(F'F)^{-1}F'$ and let $M = I - P$. Then we can write

$$S(X) = (X - F\hat{\beta})(X - F\hat{\beta})' = MX(MX)'$$

and

$$(S(X), V_t) = tr(S(X)V_t) = tr(MX(MX)'V_t) = X'MV_tMX.$$

Using the matrices $B_0 = I_n$ and $B_t = \frac{1}{2}V_t; t = 1, 2, ..., n-1$, we get the expressions

$$\hat{R}(t) = \frac{1}{n-t}X'MB_tMX; t = 0, 1, ..., n-1,$$

which are invariant quadratic forms in X, since $MF = 0$.

Using the expression for the mean value of an invariant quadratic form we get

$$E_R[\hat{R}(t)] = \frac{1}{n-t}tr(MB_tM\Sigma)$$

and from the expression for the variance of an invariant quadratic form, under the assumption that $X(.)$ is a Gaussian time series, we get

$$D_R[\hat{R}(t)] = \frac{1}{(n-t)^2}2tr((MB_tM\Sigma)^2).$$

The expressions for the mean values of $\hat{R}(t)$ can be written as

$$
\begin{aligned}
E_R[\hat{R}(t)] &= \frac{1}{n-t}tr((I-P)B_t(I-P)\Sigma) \\
&= \frac{1}{n-t}tr(B_t\Sigma - B_tP\Sigma - B_t\Sigma P + B_tP\Sigma P) \\
&= R(t) - \frac{1}{n-t}tr(B_tP\Sigma + B_t\Sigma P - B_tP\Sigma P)
\end{aligned}
$$

and we see that the estimators $\hat{R}(t)$ are not unbiased for $R(t)$. It is easy to show that in any mixed LRM there do not exist unbiased invariant quadratic estimators for $R(t); t = 0, 1, ..., n-1$.

We shall now show that the DOOLSEs $\hat{R}(t); t = 0, 1, ..., n-1$, are asymptotically unbiased. To show this we use the Schwarz inequality and the inequality $\|AB\| \leq \|A\| \|B\|$ and we get the inequalities

$$
\begin{aligned}
|tr(B_tP\Sigma)| &\leq \|B_tP\| \|\Sigma\|, \\
|tr(B_t\Sigma P)| &\leq \|PB_t\| \|\Sigma\|, \\
|tr(B_tP\Sigma P)| &\leq \|PB_tP\| \|\Sigma\| \\
&\leq \|P\| \|B_tP\| \|\Sigma\|.
\end{aligned}
$$

We shall now show that

$$\|B_t A\| \le \|A\| \quad \text{and} \quad \|A B_t\| \le \|A\|$$

for any $n \times n$ matrix A.

Since

$$B_t = \frac{1}{2}(K_t + K'_t), \quad \text{where } K_t = \begin{pmatrix} 0 & 0 \\ I_{n-t} & 0 \end{pmatrix}$$

we can write, using for A a suitable block form,

$$K_t A = \begin{pmatrix} 0 & 0 \\ I_{n-t} & 0 \end{pmatrix} \begin{pmatrix} A_{1,1} & A_{1,2} \\ A_{2,1} & A_{2,2} \end{pmatrix} = \begin{pmatrix} 0 & 0 \\ A_{1,1} & A_{1,2} \end{pmatrix}$$

from which we have the inequality

$$\|K_t A\| \le \|A\|$$

and, by analogy, we get

$$\|K'_t A\| \le \|A\|.$$

Using these inequalities we immediately get

$$\|B_t A\| = \left\| \frac{1}{2}(K_t + K'_t) A \right\| \le \frac{1}{2}(\|K_t A\| + \|K'_t A\|) \le \|A\|$$

and similarly we can get

$$\|A B_t\| \le \|A\|.$$

Thus we can write

$$\left| E_R[\hat{R}(t)] - R(t) \right| \le \frac{1}{n-t}(\|B_t P\| + \|P B_t\| + \|P\|\,\|B_t P\|)\,\|\Sigma\|$$

$$\le \frac{1}{n-t}(2\,\|P\| + \|P\|^2)\,\|\Sigma\|.$$

We remark that both P and Σ are symmetric matrices which depends on n. But it is well known that $P^2 = P$ and thus for any n:

$$\|P\|^2 = tr(P^2) = tr(P) = r(P) = k.$$

Thus we can write

$$\left| E_R[\hat{R}(t)] - R(t) \right| \le \frac{n}{n-t}(2k^{1/2} + k)\frac{\|\Sigma\|}{n},$$

$$\frac{\|\Sigma\|}{n} = \left[\frac{R^2(0)}{n} + \frac{2}{n}\sum_{t=1}^{n}\left(1 - \frac{t}{n}\right)R^2(t) \right]^{\frac{1}{2}}$$

and, as was already proved,

$$\lim_{n\to\infty} \frac{\|\Sigma\|}{n} = 0 \text{ if } \lim_{t\to\infty} R(t) = 0.$$

We have just shown that the following theorem is true.

Theorem 3.3.2. *Let $X(.)$ be a covariance stationary time series with mean values given by a LRM. Let, for its covariance function $R(.)$, the condition $\lim_{t\to\infty} R(t) = 0$ hold. Then the DOOLSEs*

$$\hat{R}_n(t) = \frac{1}{n-t}\sum_{s=1}^{n-t}(X(s+t) - \hat{m}(s+t))(X(s) - \hat{m}(s))$$

are asymptotically unbiased estimators of $R(t)$ for every fixed t.

We shall compute the variances of these estimators for a normal time series. If $X(.)$ is Gaussian, then

$$
\begin{aligned}
D_R[\hat{R}_n(t)] &= \frac{2}{(n-t)^2} tr((MB_t M\Sigma)^2) \\
&= \frac{2}{(n-t)^2} tr(((I-P)B_t(I-P)\Sigma)^2) \\
&= \frac{2}{(n-t)^2}[tr((B_t\Sigma)^2) + tr((B_t P\Sigma)^2) + \ldots + tr((PB_t P\Sigma)^2)].
\end{aligned}
$$

Using the same inequalities which hold for matrices B_t and which were used by deriving the asymptotic unbiasedness of $\hat{R}_n(t)$ we get the inequality

$$D_R[\hat{R}_n(t)] \le \frac{c(k)}{(n-t)^2}\|\Sigma\|^2,$$

where $c(k)$ is a polynomial in $k^{\frac{1}{2}}$ which does not depend on n. This is due to the fact that, for example,

$$tr((B_t\Sigma)^2) \le \|B_t\Sigma B_t\|\,\|\Sigma\| \le \|\Sigma B_t\|\,\|\Sigma\| \le \|\Sigma\|^2$$

or, by analogy,

$$tr((PB_t P\Sigma)^2) \le \|P\|^4\,\|\Sigma\|^2 \le k^2\,\|\Sigma\|^2.$$

The derived result is a base for the statement of the following theorem.

Theorem 3.3.3. *Let $X(.)$ be a Gaussian covariance stationary time series with mean values given by an LRM. Let, for its covariance function $R(.)$, the condition $\lim_{t\to\infty} R(t) = 0$ hold and let*

$$\hat{R}_n(t) = \frac{1}{n-t}\sum_{s=1}^{n-t}(X(s+t) - \hat{m}(s+t))(X(s) - \hat{m}(s))$$

be the DOOLSEs for $R(t)$. Then

$$\lim_{n\to\infty} E_R[\hat{R}_n(t) - R(t)]^2 = 0$$

for every fixed t.

Proof. This follows from the equality

$$E_R[\hat{R}_n(t) - R(t)]^2 = D_R[\hat{R}_n(t)] + (E_R[\hat{R}_n(t)] - R(t))^2,$$

from the derived inequality

$$D_R[\hat{R}_n(t)] \leq \frac{c(k)}{(n-t)^2} \|\Sigma\|^2,$$

and from the already used result according to which

$$\lim_{n\to\infty} \frac{\|\Sigma\|^2}{n^2} = 0 \text{ if } \lim_{t\to\infty} R(t) = 0$$

and from Theorem 3.3.2.

Remarks. 1. Using the *Chebyshev inequality*

$$P(\left|\hat{R}_n(t) - R(t)\right| > \varepsilon) \leq \frac{E_R[\hat{R}_n(t) - R(t)]^2}{\varepsilon^2}$$

we can prove that the DOOLSEs $\hat{R}_n(t)$ converge in probability to $R(t)$ for any fixed t under the assumptions of Theorem 3.3.3 and thus the DOOLSEs $\hat{R}_n(.)$ are consistent.

2. The sufficient conditions for consistency of the estimators $\hat{R}_n(.)$ are rather weak and they depend neither on the shape of the design matrix of the regression model, nor on the number of regression parameters.

Corollary 3.3.1. Let $\Sigma_R = \sum_{t=0}^{m-1} R(t)V_t$, where V_t are now $m \times m$ matrices, be the covariance matrix of $(X(1), ..., X(m))'$ and let

$$\hat{\Sigma}_n = \sum_{t=0}^{m-1} \hat{R}_n(t)V_t,$$

where $\hat{R}_n(t)$ are the DOOLSEs of $R(t)$ based on $(X(1), ..., X(n))'$, be an estimator of Σ_R. Then, for every fixed m, under the assumptions of Theorem 3.3.3,

$$\lim_{n\to\infty} E_R \left[\left\| \hat{\Sigma}_n - \Sigma_R \right\|^2 \right] = 0.$$

Proof. We have

$$\left\| \hat{\Sigma}_n - \Sigma_R \right\|^2 = \left\| \sum_{t=0}^{m-1} (\hat{R}_n(t) - R(t))V_t \right\|^2 = \sum_{t=0}^{m-1} (\hat{R}_n(t) - R(t))^2 \left\| V_t \right\|^2,$$

since the $m \times m$ matrices $V_t; t = 0, 1, ..., m - 1$, are orthogonal, and the proof is completed by applying the results of Theorem 3.3.3. We remark that $\|V_t\|^2$ depends only on m for every $t = 0, 1, ..., m - 1$.

In Fuller (1976) the following statement can be found on page 400. Let $X(.)$ be given by the LRM

$$X(t) = \sum_{i=1}^{k} \beta_i f_i(t) + \varepsilon(t); t \in T, \beta \in E^k,$$

where

$$\varepsilon(t) = \sum_{j=0}^{\infty} b_j Y(t - j); t \in T,$$

$\{b_j\}_{j=0}^{\infty}$ is absolutely summable and the $Y(t)$ are independent random variables with $E[Y(t)] = 0, D[Y(t)] = \sigma^2$, and with bounded $2 + \delta(\delta > 0)$ moments. Let, for the functions $f_i(.); i = 1, 2, ..., k$, the assumptions given before Theorem 3.2.2 hold and let

$$\tilde{R}_n(t) = \frac{1}{n} \sum_{s=1}^{n-t} (X(s + t) - \hat{m}(s + t))(X(s) - \hat{m}(s)); t = 0, 1, ...$$

Then the limiting behavior of

$$n^{\frac{1}{2}} [\tilde{R}_n(0) - R(0), \tilde{R}_n(1) - R(1), ..., \tilde{R}_n(1) - R(1)]$$

is, for every fixed l multivariate normal with mean value zero and with covariance matrix V, such that for the elements V_{st} of V again the equality

$$\lim_{n \to \infty} (n-t)V_{st} = (k-3)R(s)R(t) + \sum_{u=-\infty}^{\infty} [R(u)R(u-s+t) + R(u+t)R(u-s)]$$

holds for any fixed s and $t, s \geq t \geq 0$.

Untill now we have assumed that the mean value of an observed time series is given by an LRM. Now we shall study the problem of the estimation of the stationary covariance function of a time series with a mean value following an NRM. We shall use the results on estimation, of the mean value given by some NRM, which were derived in Section 3.2.

Let, as before,

$$X(t) = m_\gamma(t) + \varepsilon(t); t \in T, \gamma \in \Upsilon,$$

be a time series with mean values following a NRM and with an autoco-variance function $R(.)$. Then we can write a finite observation X of $X(.)$ as

$$X = m_\gamma + \varepsilon; \gamma \in \Upsilon, Cov(X) = \Sigma,$$

where $\Sigma_{st} = R(s-t); s, t = 1, 2, ..., n$.

The OLSE $\hat\gamma$ of γ is defined by

$$\hat\gamma(x) = \arg\min_\gamma \|x - m_\gamma\|^2 = \arg\min_\gamma \sum_{t=1}^n (x(t) - m_\gamma(t))^2; x \in \mathcal{X}.$$

It was shown that the OLSE $\hat\gamma$ can be approximated by

$$\hat\gamma_{ap}(\varepsilon) = \gamma_0 + (F'F)^{-1}F'\varepsilon + (F'F)^{-1}\left[\begin{pmatrix} \varepsilon'N_1\varepsilon \\ \vdots \\ \varepsilon'N_k\varepsilon \end{pmatrix} - \frac{1}{2}F'\begin{pmatrix} \varepsilon'A'H_1A\varepsilon \\ \vdots \\ \varepsilon'A'H_nA\varepsilon \end{pmatrix}\right].$$

In this expression we have simplified the notation by writing

$$F_{\gamma_0} = \frac{\partial m_\gamma}{\partial \gamma'}\Big|_{\gamma_0} = F, A = (F'F)^{-1}F'.$$

The $n \times n$ matrices $N_j; j = 1, 2, ..., k$, are defined by $N_j = \frac{1}{2}(O_j + O_j')$, where

$$O_{j,kl} = \sum_{t=1}^n (H_t A)_{jk} M_{tl}; k, l = 1, 2, ..., n.$$

Here $H_t; t = 1, 2, ..., n$, are the $k \times k$ Hessian matrices of m_γ defined by

$$H_{t,ij} = \frac{\partial^2 m_\gamma(t)}{\partial \gamma_i \partial \gamma_j}; i, j = 1, 2, ..., k.$$

Using a part of the Taylor expansion of m at the point $\hat\gamma$ we get

$$\hat m = m_{\hat\gamma} = m_\gamma + F(\hat\gamma - \gamma) + \frac{1}{2}(\hat\gamma - \gamma)'H_\bullet(\hat\gamma - \gamma),$$

where $(\hat\gamma - \gamma)'H_\bullet(\hat\gamma - \gamma)$ denotes the $n \times 1$ random vector with components $(\hat\gamma - \gamma)'H_t(\hat\gamma - \gamma); t = 1, 2, ..., n$. Using this notation and the notation $\varepsilon'N_\bullet\varepsilon$ for the random vector with components $\varepsilon'N_i\varepsilon; i = 1, 2, ..., k$, and $\varepsilon'A'H_\bullet A\varepsilon$ for the random vector with components $\varepsilon'A'H_tA\varepsilon; t = 1, 2, ..., n$, we can write

$$\hat\gamma_{ap}(\varepsilon) = \gamma_0 + (F'F)^{-1}F'\varepsilon + (F'F)^{-1}\left[\varepsilon'N_\bullet\varepsilon - \frac{1}{2}F'\varepsilon'A'H_\bullet A\varepsilon\right].$$

Using these new notations we can write for the residuals $\hat{\varepsilon} = X - \hat{m}$ the expression

$$
\begin{aligned}
\hat{\varepsilon} &= \varepsilon - F\left(A\varepsilon + (F'F)^{-1}\left[\varepsilon'N_\bullet\varepsilon - \frac{1}{2}F'\varepsilon'A'H_\bullet A\varepsilon\right]\right) \\
&\quad - \frac{1}{2}\left(A\varepsilon + (F'F)^{-1}\left[\varepsilon'N_\bullet\varepsilon - \frac{1}{2}F'\varepsilon'A'H_\bullet A\varepsilon\right]\right)' H_\bullet \\
&\quad \times \left(A\varepsilon + (F'F)^{-1}\left[\varepsilon'N_\bullet\varepsilon - \frac{1}{2}F'\varepsilon'A'H_\bullet A\varepsilon\right]\right).
\end{aligned}
$$

Using only the linear and quadratic, in components of ε, terms we can approximate the residuals $\hat{\varepsilon}$ by $\hat{\varepsilon}_{ap}$ given by

$$
\hat{\varepsilon}_{ap} = M\varepsilon - A'\varepsilon'N_\bullet\varepsilon - \frac{1}{2}M\varepsilon'A'H_\bullet A\varepsilon.
$$

These residuals will be used for an unknown covariance function $R(.)$. Some properties of residuals for the case of uncorrelated errors were studied by Cook and Tsai (1985).

Now let us consider the random matrix $\hat{\Sigma}_{ap}$ given by

$$
\begin{aligned}
\hat{\Sigma}_{ap} &= \hat{\varepsilon}_{ap}\hat{\varepsilon}'_{ap} = M\varepsilon\varepsilon'M - A'\varepsilon'N_\bullet\varepsilon M - M\varepsilon\varepsilon'N_\bullet\varepsilon M + A'\varepsilon'N_\bullet\varepsilon(\varepsilon'N_\bullet\varepsilon)'A \\
&\quad - \frac{1}{2}M\varepsilon(\varepsilon'A'H_\bullet A\varepsilon)'M - \frac{1}{2}M\varepsilon'A'H_\bullet A\varepsilon\varepsilon'M \\
&\quad + \frac{1}{2}A'\varepsilon'N_\bullet\varepsilon(\varepsilon'A'H_\bullet A\varepsilon)'M + \frac{1}{2}M\varepsilon'A'H_\bullet A\varepsilon(\varepsilon'N_\bullet\varepsilon)'M \\
&\quad + \frac{1}{4}M\varepsilon'A'H_\bullet A\varepsilon(\varepsilon'A'H_\bullet A\varepsilon)'M.
\end{aligned}
$$

The estimators $\hat{R}(t)$ of $R(t); t = 0, 1, 2, ..., n-1$, given by

$$
\hat{R}(t) = \frac{1}{n-t}\sum_{s=1}^{n-t}(X(s+t) - \hat{m}(s+t))(X(s) - \hat{m}(s)); t = 0, 1, ..., n-1,
$$

are the natural generalizations of the DOOLSEs of $R(.)$ for the case when the mean value follows an LRM which has already been studied. Thus these estimators can again be called the DOOLSEs. They can be approximated by the estimators

$$
\hat{R}_{ap}(t) = \frac{1}{n-t}\sum_{s=1}^{n-t}\hat{\varepsilon}_{ap}(s+t)\hat{\varepsilon}_{ap}(s).
$$

Again, as in the case with an LRM, we can write

$$\hat{R}_{ap}(t) = \frac{1}{n-t} tr(B_t \hat{\Sigma}_{ap}); t = 0, 1, ..., n - 1,$$

where, as before,

$$B_t = \frac{1}{2}\left[\begin{pmatrix} 0 & I_{n-t} \\ 0 & 0 \end{pmatrix} + \begin{pmatrix} 0 & 0 \\ I_{n-t} & 0 \end{pmatrix}\right]; t = 0, 1, ..., n - 1,$$

and thus all the inequalities derived for B_t can also be used in the case of an NRM.

We can also write

$$\hat{R}_{ap}(t) = \frac{1}{n-t}[\varepsilon' M B_t M \varepsilon - 2\varepsilon' M B_t A' \varepsilon' N_\bullet \varepsilon - (\varepsilon' N_\bullet \varepsilon)' A B_t A' \varepsilon' N_\bullet \varepsilon$$
$$- (\varepsilon' A' H_\bullet A\varepsilon)' M B_t M \varepsilon + (\varepsilon' A' H_\bullet A\varepsilon)' M B_t A' \varepsilon' N_\bullet \varepsilon$$
$$+ \frac{1}{4}(\varepsilon' A' H_\bullet A\varepsilon)' M B_t M \varepsilon' A' H_\bullet A\varepsilon].$$

Now we shall study the limit properties, as n tends to infinity, of the estimators $\hat{R}_{ap}(.)$. We remark that the matrices Σ, M, H and others, and also their norms depend on n, but this will not be announced later on.

Theorem 3.3.4. *Let, in the NRM,*

$$(F_n' F_n)^{-1} = \frac{1}{n} G_n,$$

where $\lim\limits_{n\to\infty} G_n = G$ *and* G *is a n.d. matrix. Next, let the following limits*

$$\lim_{n\to\infty} \frac{1}{n} \sum_{t=1}^{n} \frac{\partial m_\gamma(t)}{\partial \gamma_i} \frac{\partial^2 m_\gamma(t)}{\partial \gamma_j \partial \gamma_k},$$

$$\lim_{n\to\infty} \frac{1}{n} \sum_{t=1}^{n} \frac{\partial m_\gamma^2(t)}{\partial \gamma_i \partial \gamma_j} \frac{\partial^2 m_\gamma(t)}{\partial \gamma_k \partial \gamma_l}$$

exist and be finite for every i, j, k, l. *Let the errors* ε *have* $N_n(0, \Sigma)$ *distribution with* $\Sigma_{ij} = R(i - j)$ *and let*

$$\lim_{n\to\infty} \frac{1}{n} \|\Sigma\| = 0.$$

Then the estimators

$$\hat{R}_{ap}(t) = \frac{1}{n-t} \sum_{s=1}^{n-t} \hat{\varepsilon}_{ap}(s+t)\hat{\varepsilon}_{ap}(s)$$

converge for every fixed t in probability to $R(t)$ as n tends to infinity.

Proof. It was shown in Theorem 3.3.3 that

$$\lim_{n \to \infty} E\left[\frac{1}{n}\varepsilon' M B_t M\varepsilon - R(t)\right]^2 = 0$$

if $\lim_{n \to \infty} 1/n \|\Sigma\| = 0$ and thus $1/n\ \varepsilon' M B_t M\varepsilon$ converges in probability to $R(t)$. Thus the theorem will be proved if we show that all the other members appearing in the expression for $\hat{R}_{ap}(t)$ converge in probability to zero. Let us consider the term $1/n\ \varepsilon' M B_t A' \varepsilon' N_\bullet \varepsilon$. We can write

$$\left|\frac{1}{n}\varepsilon' M B_t A' \varepsilon' N_\bullet\varepsilon\right|^2 \leq \frac{1}{n^2}\|M\varepsilon\|^2 \|B_t A' \varepsilon' N_\bullet\varepsilon\|^2$$

$$\leq \frac{1}{n}\varepsilon' M\varepsilon\frac{1}{n}\|A' \varepsilon' N_\bullet\varepsilon\|^2.$$

Now we shall prove that $1/n\ \varepsilon' M\varepsilon$ converges in probability to $R(0)$ and $1/n\ \|A' \varepsilon' N_\bullet\varepsilon\|^2$ converges in probability to zero and thus their product converges in probability to zero. But

$$\frac{1}{n}\varepsilon' M\varepsilon = \frac{1}{n}\varepsilon' M B_0 M\varepsilon,$$

since $B_0 = I_n$ and $M^2 = M$, and it was already shown that this term converges to $R(0)$. Next we have

$$(A' \varepsilon' N_\bullet\varepsilon)_i = \sum_{t=1}^n (A' H_t A\varepsilon)_i (M\varepsilon)_t; i = 1, 2, ..., n,$$

and

$$\frac{1}{n}\|A' \varepsilon' N_\bullet\varepsilon\|^2 = \frac{1}{n}\sum_{i=1}^n\left(\sum_{t=1}^n (A' H_t A\varepsilon)_i (M\varepsilon)_t\right)^2$$

$$\leq \frac{1}{n}\sum_{i=1}^n\sum_{t=1}^n (A' H_t A\varepsilon)_i^2 \sum_{t=1}^n (M\varepsilon)_t^2$$

$$= \sum_{t=1}^n \varepsilon' A' H_t A A' H_t A\varepsilon\frac{1}{n}\varepsilon' M\varepsilon.$$

Thus it is sufficient to prove that $\sum_{t=1}^n \varepsilon' A' H_t A' H_t A\varepsilon$ converges in probability to zero. But

$$E\left[\sum_{t=1}^n \varepsilon' A' H_t A A' H_t A\varepsilon\right] = \sum_{t=1}^n tr(A' H_t A A' H_t A\Sigma)$$

and thus

$$\left| E\left[\sum_{t=1}^{n} \varepsilon' A' H_t A A' H_t A\varepsilon \right] \right| \leq \sum_{t=1}^{n} |tr(A' H_t A A' H_t A\Sigma)|$$

$$\leq \sum_{t=1}^{n} \|A' H_t A\|^2 \|\Sigma\|.$$

Next,

$$\sum_{t=1}^{n} \|A' H_t A\|^2 = \sum_{t=1}^{n} tr(A' H_t A A' H_t A)$$

$$= tr\left(\sum_{t=1}^{n} H_t (F'F)^{-1} H_t (F'F)^{-1} \right),$$

since $AA' = (F'F)^{-1}$. Thus we have

$$\lim_{n\to\infty} E\left[\sum_{t=1}^{n} \varepsilon' A' H_t A A' H_t A\varepsilon \right] = 0$$

if the assumptions of the theorem are fulfilled. From the same reasons, using the expression $D[\varepsilon' C\varepsilon] = 2tr(C\Sigma C\Sigma)$, we get

$$\lim_{n\to\infty} D\left[\sum_{t=1}^{n} \varepsilon' A' H_t A A' H_t A\varepsilon \right] = 0.$$

Next,

$$\frac{1}{n} |(\varepsilon' N.\varepsilon)' A B_t A' \varepsilon' N.\varepsilon| \leq \frac{1}{n} \|A' \varepsilon' N.\varepsilon\|^2$$

and the last term converges in probability to zero as we have just shown. Further we can write, since $M = I - P$,

$$\frac{1}{n} |(\varepsilon' A' H_\bullet A\varepsilon)' M B_t M\varepsilon| = \frac{1}{n} |(\varepsilon' A' H_\bullet A\varepsilon)'(I - P) B_t M\varepsilon|$$

$$\leq \frac{1}{n} |(\varepsilon' A' H_\bullet A\varepsilon)' B_t M\varepsilon|$$

$$+ \frac{1}{n} |(\varepsilon' A' H_\bullet A\varepsilon)' P B_t M\varepsilon|$$

and

$$\frac{1}{n^2} |(\varepsilon' A' H_\bullet A\varepsilon)' B_t M\varepsilon|^2 \leq \frac{1}{n} \|\varepsilon' A' H_\bullet A\varepsilon\|^2 \frac{1}{n} \varepsilon' M\varepsilon$$

$$\leq \varepsilon' A' A\varepsilon \frac{1}{n} \sum_{t=1}^{n} \varepsilon' A' H_t^2 A\varepsilon \frac{1}{n} \varepsilon' M\varepsilon.$$

It is easy to prove that the mean values and variances of $\varepsilon' A' A \varepsilon$ and $1/n$ $\sum_{t=1}^{n} \varepsilon' A' H_t^2 A \varepsilon$ converge to zero under the assumptions of the theorem and thus these random variables converge in probability to zero. Next we get, under the assumption that $r(P) = k$,

$$
\begin{aligned}
\frac{1}{n^2} |(\varepsilon' A' H_\bullet A \varepsilon)' P B_t M \varepsilon|^2 &\leq \frac{1}{n^2} \|\varepsilon' A' H_\bullet A \varepsilon\|^2 \|P B_t M \varepsilon\|^2 \\
&\leq \frac{1}{n^2} \|\varepsilon' A' H_\bullet A \varepsilon\|^2 \|P\|^2 \|B_t M \varepsilon\|^2 \\
&\leq \frac{k}{n} \|\varepsilon' A' H_\bullet A \varepsilon\|^2 \frac{1}{n} \varepsilon' M \varepsilon.
\end{aligned}
$$

Let us consider the last term in the expression for $\hat{R}_{ap}(.)$. We get, as before,

$$
|(\varepsilon' A' H_\bullet A \varepsilon)' M B_t A' \varepsilon' N \varepsilon| \leq |(\varepsilon' A' H_\bullet A \varepsilon)' B_t A' \varepsilon' N \varepsilon| \\
+ |(\varepsilon' A' H_\bullet A \varepsilon)' P B_t A' \varepsilon' N \varepsilon|.
$$

Next,

$$
\frac{1}{n^2} |(\varepsilon' A' H_\bullet A \varepsilon)' B_t A' \varepsilon' N \varepsilon|^2 \leq \frac{1}{n} \|\varepsilon' A' H_\bullet A \varepsilon\|^2 \frac{1}{n} \|A' \varepsilon' N \varepsilon\|^2
$$

and we know from our proof that both terms on the right-hand side of the last inequality converge to zero. Finally,

$$
\frac{1}{n^2} |(\varepsilon' A' H_\bullet A \varepsilon)' P B_t A' \varepsilon' N \varepsilon|^2 \leq \frac{k}{n} \|\varepsilon' A' H_\bullet A \varepsilon\|^2 \frac{1}{n} \|A' \varepsilon' N \varepsilon\|^2
$$

and this term can be bounded from above in the same way.

The proof of the theorem now follows from the derived results and from the well-known facts on convergence in probability:

(a) $X_n \to X$ iff $X_n^2 \to X^2$;

(b) if $X_n \to X$ and $Y_n \to Y$, then $X_n Y_n \to XY$ and $aX_n + bY_n \to aX + bY$;

(c) if $|X_n| \leq |Y_n|$ and $Y_n \to 0$, then $X_n \to 0$; and

d) if $E[X_n] \to 0$ and $D[X_n] \to 0$, then $X_n \to 0$.

Remarks. 1. The conditions of the theorem required for the mean value $m_\gamma(.)$ are similar to those appearing in Jennrich (1969), Wu (1981) and others studying the limit properties of the OLSE $\hat{\gamma}$ of γ. As we have shown, for the consistency of estimators of the stationary covariance function weaker conditions than for consistency of regression parameters are required, if the regression model is linear. A similar situation occurs in the case of nonlinear regression.

2. For estimating $R(0)$ we have $B_0 = I_n$. Two terms in the expression for $\hat{R}_{ap}(0)$ vanish in this case, since $MA' = 0$.

3. If the errors are uncorrelated with a common variance σ^2, then the condition of the theorem on the covariance matrix Σ is fulfilled. In this case $1/n \, \|\Sigma\| = n^{-\frac{1}{2}}\sigma$.

4. For stationary errors, as before, the condition $\lim_{n\to\infty} 1/n \, \|\Sigma\| = 0$ can be replaced by the more natural condition $\lim_{t\to\infty} R(t) = 0$.

Example 3.3.1. Let us consider a time series $X(.)$ following the NRM

$$X(t) = \gamma_1 + \gamma_2 t + \sum_{i=1}^{2}(\beta_i^{(1)} \cos \lambda_i t + \beta_i^{(2)} \sin \lambda_i t) + \varepsilon(t); t \in T,$$

where $\gamma = (\gamma_1, \gamma_2, \beta_1^{(1)}, \beta_1^{(2)}, \beta_2^{(1)}, \beta_2^{(2)}, \lambda_1, \lambda_2)'$ is an unknown vector of regression parameters and $\varepsilon(.)$ is an AR(1) time series given by

$$\varepsilon(t) = \rho\varepsilon(t-1) + e(t)$$

with a white noise $e(.)$ having variance σ^2 and with an autoregression parameter ρ. The problem of the estimation of parameter γ was studied in Example 3.2.7. Now we shall illustrate properties of the DOOLSEs $\hat{R}(.)$.

We have simulated three realizations of $X(.)$ of different lengths with $\gamma = (3, 2, 3, 2, 3, 4, .75, .25)', \sigma^2 = 1$, and with different values of the autoregression parameter ρ. In the following tables corresponding values of $\hat{R}(.)$ and, for comparison, also values of estimates $\hat{R}_\varepsilon(.)$ computed from realizations of the AR(1) time series $\varepsilon(.)$ with the mean value zero are given.

Table 3.4. Estimates of $R(.)$ for $n = 51$.

t	$\hat{R}(.)$ $\rho = -.8$	$\hat{R}_\varepsilon(.)$	$\hat{R}(.)$ $\rho = 0$	$\hat{R}_\varepsilon(.)$	$\hat{R}(.)$ $\rho = .4$	$\hat{R}_\varepsilon(.)$	$\hat{R}(.)$ $\rho = .8$	$\hat{R}_\varepsilon(.)$
0	3.1	3.2	0.7	1.0	0.8	1.1	1.1	1.7
1	-2.6	-2.6	-0.1	0.0	0.3	0.4	0.7	1.2
2	2.2	2.3	0.0	0.0	0.0	0.0	0.4	0.7
3	-2.0	-2.0	-0.1	-0.1	-0.1	-0.2	0.2	0.2
4	1.9	1.8	0.1	-0.1	0.0	-0.2	0.0	0.0
5	-1.8	-1.8	-0.2	-0.2	-0.2	-0.3	-0.3	-0.3

Table 3.5. Estimates of $R(.)$ for $n = 101$.

t	$\hat{R}(.)$ $\rho = -.8$	$\hat{R}_\varepsilon(.)$	$\hat{R}(.)$ $\rho = 0$	$\hat{R}_\varepsilon(.)$	$\hat{R}(.)$ $\rho = .4$	$\hat{R}_\varepsilon(.)$	$\hat{R}(.)$ $\rho = .8$	$\hat{R}_\varepsilon(.)$
0	2.5	2.6	0.9	1.0	0.9	0.9	1.7	1.8
1	-2.0	-2.1	0.0	0.0	0.4	0.4	1.1	1.2
2	1.8	1.8	-0.1	-0.1	0.0	0.0	0.6	0.6
3	-1.7	-1.7	-0.1	-0.1	-0.1	-0.1	0.4	0.3
4	1.4	1.4	0.0	-0.1	-0.1	-0.2	0.1	0.0
5	-1.1	-1.1	-0.1	-0.1	-0.2	-0.2	0.0	-0.1

Table 3.6. Estimates of $R(.)$ for $n = 201$.

	$\hat{R}(.)$	$\hat{R}_\varepsilon(.)$	$\hat{R}(.)$	$\hat{R}_\varepsilon(.)$	$\hat{R}(.)$	$\hat{R}_\varepsilon(.)$	$\hat{R}(.)$	$\hat{R}_\varepsilon(.)$
t	$\rho = -.8$		$\rho = 0$		$\rho = .4$		$\rho = .8$	
0	2.1	2.1	1.0	1.0	1.1	1.3	2.1	2.1
1	-1.5	-1.5	0.0	0.0	0.5	0.5	1.6	1.6
2	1.1	1.1	-0.1	-0.1	0.0	0.0	1.0	1.0
3	-0.8	-0.8	0.0	0.0	0.0	0.0	0.6	0.6
4	0.8	0.6	-0.1	-0.1	-0.1	-0.1	0.4	0.3
5	-0.6	-0.6	0.0	0.0	-0.1	-0.1	0.0	0.0

It can be seen from these tables that the influence of an unknown mean value, following the NRM with an eight dimensional vector of regression parameters, on estimation of a covariance function, is not very large for relatively small $n = 51$. For $n = 101$ and $n = 201$ the influence of the mean value is negligible for all ρ's.

3.4   Maximum Likelihood Estimation

Before starting to solve the problem of the maximum likelihood estimation of the parameters of time series we shall generalize the notion of doubly least squares estimators. This notion was introduced in Chapter 1 for the case when mean values and covariance matrices of an observed vector are given by LRMs. This notion can also be used in the case when either mean values or covariance matrices, or both of these characteristics, are given by NRMs.

Let us assume that we have an observation X of time series $X(.)$ given by an additive regression model

$$X = m + \varepsilon; E[\varepsilon] = 0, m \in \mathcal{M}, Cov(X) = \Sigma \in \Xi,$$

where

$$\mathcal{M} = \{m_\beta = F\beta; \beta \in E^k\}$$

for a LRM, or

$$\mathcal{M} = \{m_\gamma; \gamma \in \Gamma\}$$

for a NRM, and

$$\Xi = \{\Sigma_\nu; \nu \in \Upsilon\}.$$

Let us denote by $\hat{m}$ the vector $m_{\hat{\beta}}$ or the vector $m_{\hat{\gamma}}$, where $\hat{\beta}$ and $\hat{\gamma}$ are the OLSEs of β and γ, respectively, and let

$$S(X) = (X - \hat{m})(X - \hat{m})'.$$

Then the estimator $\hat{\nu}$ defined by

$$\hat{\nu}(X) = \arg\min_{\nu} \|S(X) - \Sigma_\nu\|^2$$

will be called the DOOLSE of ν.

In the case when $\Sigma_\nu; \nu \in \Upsilon$, are given by an linear regression model we have an explicit expression for the DOOLSE, which is given in Example 1.4.3.

If covariance matrices $\Sigma_\nu; \nu \in \Upsilon$, depend on parameter ν nonlinearly, then $\|S(X) - \Sigma_\nu\|^2; \nu \in \Upsilon$, is a nonlinear function defined on Υ and we must use some iterative procedure for computing $\hat{\nu}(x); x \in \mathcal{X}$.

Now let $\hat{m}_\nu$ denote the random vector $F\hat{\beta}_\nu$, or the random vector $m_{\hat{\gamma}_\nu}$, where $\hat{\beta}_\nu$ and $\hat{\gamma}_\nu$ are the WELSEs of β and γ, respectively, defined by

$$\hat{\gamma}_\nu(X) = \arg\min_{\gamma} \|X - m_\gamma\|^2_{\Sigma_\nu^{-1}}$$

and $\hat{\beta}_\nu$ by the same expression but with γ replaced by β. It is clear that $\hat{\beta}_\nu$ is equal to the BLUE β^*_ν of β.

Let

$$S_\nu(X) = (X - \hat{m}_\nu)(X - \hat{m}_\nu)'.$$

Then the estimator $\hat{\nu}_\nu$ defined by

$$\hat{\nu}_\nu(X) = \arg\min_{\nu} \|S_\nu(X) - \Sigma_\nu\|^2_{\Sigma_\nu^{-1}}$$

will be called the DOWELSE of ν.

It should be noted that even in the case when the mean values and covariances of X are given by LRMs the function $g(\nu) = \|S_\nu(X) - \Sigma_\nu\|^2_{\Sigma_\nu^{-1}}$; $\nu \in \Upsilon$, is a nonlinear function of ν and thus its minimum should be computed by using some iterative method.

Let us now assume that the observation X of a time series $X(.)$ is given by an LRM

$$X = F\beta + \varepsilon; \beta \in E^k, E[\varepsilon] = 0, Cov(X) \in \Xi = \{\Sigma_\nu; \nu \in \Upsilon\},$$

and let $X \sim N_n(F\beta, \Sigma_\nu)$. Then the likelihood function L is given by

$$L_x(\beta, \nu) = \frac{1}{(2\pi)^{n/2} (\det \Sigma_\nu)^{1/2}} \exp\{-\frac{1}{2}(x - F\beta)'\Sigma_\nu^{-1}(x - F\beta)\}$$

and the loglikelihood function by

$$\ln L_x(\beta,\nu) = -\frac{n}{2}\ln 2\pi - \frac{1}{2}\ln\det(\Sigma_\nu) - \frac{1}{2}\|x - F\beta\|^2_{\Sigma_\nu^{-1}}.$$

The MLEs $\tilde{\beta}(x), \tilde{\nu}(x)$ for β, ν are given by

$$
\begin{aligned}
(\tilde{\beta}(x), \tilde{\nu}(x))' &= \arg\max_\nu \max_{\beta|\Sigma_\nu} \ln L_x(\beta,\nu) \\
&= \arg\max_\nu [-\frac{1}{2}\ln\det(\Sigma_\nu) + \max_{\beta|\Sigma_\nu} -\frac{1}{2}\|x - F\beta\|^2_{\Sigma_\nu^{-1}}].
\end{aligned}
$$

But for a given ν we have

$$\arg\max_{\beta|\Sigma_\nu}[-\frac{1}{2}\|x - F\beta\|^2_{\Sigma_\nu^{-1}}] = \arg\min_{\beta|\Sigma_\nu}\|x - F\beta\|^2_{\Sigma_\nu^{-1}}$$

and we know that

$$\min_{\beta|\Sigma_\nu}\|x - F\beta\|^2_{\Sigma_\nu^{-1}} = \|x - F\beta_\nu^*(x)\|^2_{\Sigma_\nu^{-1}},$$

where

$$\beta_\nu^*(x) = (F'\Sigma_\nu^{-1}F)^{-1}F'\Sigma_\nu^{-1}x$$

is the BLUE of β. Next we can write

$$
\begin{aligned}
\tilde{\nu}(x) &= \arg\max_\nu[-\frac{1}{2}\ln\det(\Sigma_\nu) - \frac{1}{2}\|x - F\beta_\nu^*(x)\|^2_{\Sigma_\nu^{-1}}], \\
\tilde{\beta}(x) &= \beta_{\tilde{\nu}(x)}^*(x) = (F'\Sigma_{\tilde{\nu}(x)}^{-1}F)^{-1}F'\Sigma_{\tilde{\nu}(x)}^{-1}x.
\end{aligned}
$$

In many cases the MLEs $\tilde{\beta}$ and $\tilde{\nu}$ for unknown parameters β and ν are a solution of the likelihood equations

$$\frac{\partial}{\partial\beta}\ln L_x(\beta,\nu) \Big|_{\tilde{\beta},\tilde{\nu}} = 0,$$

$$\frac{\partial}{\partial\nu}\ln L_x(\beta,\nu) \Big|_{\tilde{\beta},\tilde{\nu}} = 0.$$

For computing $\partial/\partial\nu \ln L_x(\beta,\nu)$ we use the well-known formulas

$$\frac{\partial}{\partial\nu}\ln\det(\Sigma_\nu) = tr\left(\Sigma_\nu^{-1}\frac{\partial\Sigma_\nu}{\partial\nu}\right)$$

and

$$\frac{\partial\Sigma_\nu^{-1}}{\partial\nu} = -\Sigma_\nu^{-1}\frac{\partial\Sigma_\nu}{\partial\nu}\Sigma_\nu^{-1}.$$

Using these expressions we get the likelihood equations

$$(F'\Sigma_\nu^{-1}F)\beta - F'\Sigma_\nu^{-1}x = 0,$$

$$tr\left(\Sigma_\nu^{-1}\frac{\partial\Sigma_\nu}{\partial\nu}\right) - (x - F\beta)'\,\Sigma_\nu^{-1}\frac{\partial\Sigma_\nu}{\partial\nu}\Sigma_\nu^{-1}(x - F\beta) = 0.$$

The first equation has, for every $\nu \in \Upsilon$ and $x \in \mathcal{X}$, the solution

$$\tilde{\beta}_\nu(x) = (F'\Sigma_\nu^{-1}F)^{-1}F'\Sigma_\nu^{-1}x$$

and the MLE $\tilde{\nu}(x); x \in \mathcal{X}$, is a solution of the equation

$$tr\left(\Sigma_\nu^{-1}\frac{\partial\Sigma_\nu}{\partial\nu}\right) = (x - F\tilde{\beta}_\nu(x))'\,\Sigma_\nu^{-1}\frac{\partial\Sigma_\nu}{\partial\nu}\Sigma_\nu^{-1}(x - F\tilde{\beta}_\nu(x)).$$

The solution $\tilde{\nu}(X)$ of the last nonlinear equation could be found by using some iterative method.

Example 3.4.1. Let us consider an observation X of a Gaussian AR(1) time series

$$X(t) = \rho X(t - 1) + \varepsilon(t); t = 1, 2, ...,$$

with $\nu = (\rho, \sigma^2)'$, where $\rho \in (-1, 1)$ and $\sigma^2 \in (0, \infty)$. Then it is well known that $E[X(t)] = 0$ and

$$\Sigma_{\nu,st} = R_\nu(s - t) = \frac{\sigma^2}{1 - \rho^2}\rho^{|s-t|}; s, t = 1, 2, ..., n.$$

The likelihood equations in this case are

$$tr\left(\Sigma_\nu^{-1}\frac{\partial\Sigma_\nu}{\partial\rho}\right) = x'\Sigma_\nu^{-1}\frac{\partial\Sigma_\nu}{\partial\rho}\Sigma_\nu^{-1}x$$

and

$$tr\left(\Sigma_\nu^{-1}\frac{\partial\Sigma_\nu}{\partial\sigma^2}\right) = x'\Sigma_\nu^{-1}\frac{\partial\Sigma_\nu}{\partial\sigma^2}\Sigma_\nu^{-1}x.$$

Next, we know that

$$\Sigma_\nu = \frac{\sigma^2}{1 - \rho^2}\begin{pmatrix} 1 & \rho & \cdots & \rho^{n-2} & \rho^{n-1} \\ \rho & 1 & \cdots & \rho^{n-3} & \rho^{n-2} \\ \cdot & \cdot & \cdots & \cdot & \cdot \\ \rho^{n-2} & \rho^{n-1} & \cdots & 1 & \rho \\ \rho^{n-1} & \rho^{n-2} & \cdots & \rho & 1 \end{pmatrix},$$

$$\Sigma_\nu^{-1} = \frac{1}{\sigma^2}U_\rho,$$

where

$$U_\rho = \begin{pmatrix} 1 & -\rho & 0 & \cdots & 0 & 0 & 0 \\ -\rho & 1+\rho^2 & -\rho & \cdots & 0 & 0 & 0 \\ \cdot & \cdot & \cdot & \cdots & \cdot & \cdot & \cdot \\ 0 & 0 & 0 & \cdots & -\rho & 1+\rho^2 & -\rho \\ 0 & 0 & 0 & \cdots & 0 & -\rho & 1 \end{pmatrix}.$$

Thus we can write

$$\frac{\partial \Sigma_\nu}{\partial \rho} = \frac{2\rho}{1-\rho^2}\Sigma_\nu + \frac{\sigma^2}{1-\rho^2}V_\rho,$$

where the matrix V_ρ is given by

$$V_\rho = \begin{pmatrix} 0 & 1 & \cdots & (n-2)\rho^{n-3} & (n-1)\rho^{n-2} \\ 1 & 0 & \cdots & (n-3)\rho^{n-4} & (n-2)\rho^{n-3} \\ \cdot & \cdot & \cdots & \cdot & \cdot \\ (n-2)\rho^{n-3} & (n-1)\rho^{n-2} & \cdots & 0 & 1 \\ (n-1)\rho^{n-2} & (n-2)\rho^{n-3} & \cdots & 1 & 0 \end{pmatrix}$$

and

$$\frac{\partial \Sigma_\nu}{\partial \sigma^2} = \frac{1}{\sigma^2}\Sigma_\nu.$$

Thus the likelihood equations are

$$tr\left(\frac{2\rho}{1-\rho^2}I_n + \frac{\sigma^2}{1-\rho^2}\Sigma_\nu^{-1}V_\rho\right) = x'\Sigma_\nu^{-1}\left(\frac{2\rho}{1-\rho^2}\Sigma_\nu + \frac{\sigma^2}{1-\rho^2}V_\rho\right)\Sigma_\nu^{-1}x$$

and

$$tr\left(\frac{1}{\sigma^2}I_n\right) = \frac{1}{\sigma^2}x'\Sigma_\nu^{-1}x,$$

or

$$2\rho = 2\rho x'\Sigma_\nu^{-1}x + \sigma^2 x'\Sigma_\nu^{-1}V_\rho\Sigma_\nu^{-1}x$$

and

$$n = x'\Sigma_\nu^{-1}x.$$

The likelihood equations can be written as

$$(n-1)\rho = -\frac{1}{2\sigma^2}x'U_\rho V_\rho U_\rho x,$$

$$n = x'\Sigma_\nu^{-1}x.$$

From the last equation we get the expression for the MLE $\tilde{\sigma}^2$ of σ^2 :

$$\tilde{\sigma}^2(x) = \frac{1}{n}\sum_{t=1}^{n} x^2(t) + \tilde{\rho}^2(x)\frac{1}{n}\sum_{t=2}^{n-1} x^2(t) - 2\tilde{\rho}(x)\frac{1}{n}\sum_{t=2}^{n} x(t)x(t-1)$$

which also contains the MLE $\tilde{\rho}$ of ρ. The likelihood equation

$$2\tilde{\sigma}^2(x)(n-1)\rho = -x'U_\rho V_\rho U_\rho x$$

for the MLE $\tilde{\rho}(x)$ of ρ is a third degree polynomial in ρ. It is shown in Beach and MacKinnon (1978) that this polynomial has one real root $\tilde{\rho}(x)$ belonging to $(-1, 1)$.

There is also another approach to the problem of maximum likelihood estimation in AR time series. For dependent observations the likelihood function $L_x(\nu)$ can be written as

$$L_x(\nu) = \prod_{t=1}^{n} f_\nu(x(t) \mid x(1), ..., x(t-1)),$$

where $f_\nu(x(1) \mid x(0)) = f_\nu(x(1))$. Since we assume that

$$X(t) = \rho X(t-1) + \varepsilon(t); t = 1, 2, ...,$$

we can write $X(t) \mid x(t-1) \sim N(\rho x(t-1), \sigma^2)$ and

$$L_x(\nu) = -\frac{n-1}{2}\ln 2\pi - \frac{n-1}{2}\ln\sigma^2 + \frac{1}{2\sigma^2}\sum_{t=2}^{n}(x(t)-\rho x(t-1))^2 + \ln f_\nu(x(1)).$$

If we assume that $|\rho| < 1$, then $X(1) \sim N(0, \sigma^2/(1-\rho^2))$, the time series $X(.)$ is stationary, and

$$L_x(\nu) = -\frac{n}{2}\ln 2\pi - \frac{n}{2}\ln\sigma^2 - \frac{1-\rho^2}{2\sigma^2}x^2(1) + \frac{1}{2\sigma^2}\sum_{t=2}^{n}(x(t) - \rho x(t-1))^2.$$

The likelihood equations are again nonlinear and to compute the MLE $\tilde{\nu} = (\tilde{\rho}, \tilde{\sigma}^2)$ we again have to find the real root of the third-order polynomial in ρ.

In many practical applications it is assumed that $X(1)$ is not random and that it is equal to some $x(1)$. Then the modified likelihood is

$$L_x(\nu) = -\frac{n-1}{2}\ln 2\pi - \frac{n-1}{2}\ln\sigma^2 + \frac{1}{2\sigma^2}\sum_{t=2}^{n}(x(t) - \rho x(t-1))^2$$

and the *modified least squares estimators* $\hat{\rho}$ and $\hat{\sigma}^2$ are given by

$$\hat{\rho}(x) = \frac{\sum\limits_{t=2}^{n} x(t)x(t-1)}{\sum\limits_{t=1}^{n} x^2(t)}$$

and

$$\hat{\sigma}^2(x) = \frac{1}{n-1} \sum_{t=2}^{n} (x(t) - \hat{\rho}(x)x(t-1))^2.$$

The estimator $\hat{\rho}$ is, in the economic literature, called *the serial coefficient of correlation*. It is also called *the least squares estimator*, since

$$\hat{\rho}(x) = \arg\min_{\rho} \sum_{t=2}^{n} (x(t) - \rho x(t-1))^2.$$

Let us now consider the case when the observation X of time series $X(.)$ is given by a mixed LRM

$$X = F\beta + \varepsilon; \beta \in E^k, E[\varepsilon] = 0,$$

$$Cov(X) \in \Xi = \left\{ \Sigma_\nu : \Sigma_\nu = \sum_{j=1}^{l} \nu_j V_j, ; \nu \in \Upsilon \right\}.$$

In this case the last likelihood equations can be written in the form

$$tr\left(\Sigma_\nu^{-1} \frac{\partial \Sigma_\nu}{\partial \nu_i} \Sigma_\nu^{-1} \Sigma_\nu \right) = tr\left(\Sigma_\nu^{-1} \frac{\partial \Sigma_\nu}{\partial \nu_i} \Sigma_\nu^{-1} (x - F\beta_\nu^*(x))(x - F\beta_\nu^*(x))' \right)$$

for $i = 1, 2, ..., l$, where

$$\beta_\nu^*(x) = (F'\Sigma_\nu^{-1}F)^{-1}F'\Sigma_\nu^{-1}x,$$

or, equivalently,

$$tr\left(\Sigma_\nu^{-1} V_i \Sigma_\nu^{-1} \sum_{j=1}^{l} \nu_j V_j \right) = tr(\Sigma_\nu^{-1} V_i \Sigma_\nu^{-1} S_\nu(x)); i = 1, 2, ..., l,$$

where

$$S_\nu(x) = (x - F\beta_\nu^*(x))(x - F\beta_\nu^*(x))'.$$

From the last equation we easily get

$$\sum_{j=1}^{l} tr(V_i\Sigma_\nu^{-1}V_j\Sigma_\nu^{-1})\nu_j = tr(V_i\Sigma_\nu^{-1}S_\nu(x)\Sigma_\nu^{-1}); i = 1, 2, ..., l,$$

or, in the matrix form,

$$G_\nu\nu = g(x, \nu),$$

where G_ν is the $l \times l$ matrix with elements

$$G_{\nu,ij} = tr(V_i\Sigma_\nu^{-1}V_j\Sigma_\nu^{-1}) = (V_i, V_j)_{\Sigma_\nu^{-1}}; i, j = 1, 2, ..., l,$$

and $g(x, \nu)$ is a $l \times 1$ vector with components

$$g(x, \nu)_i = tr(V_i\Sigma_\nu^{-1}S_\nu(x)\Sigma_\nu^{-1}) = (V_i, S_\nu(x))_{\Sigma_\nu^{-1}}; i = 1, 2, ..., l.$$

We can use the new notation

$$g(x, \nu) = (V, S_\nu(x))_{\Sigma_\nu^{-1}}.$$

Remark. It follows from the projection theory that, for any given value ν_0 of a parameter ν, a solution $\nu^{(0)}(x)$ of the equation

$$G_{\nu_0}\nu = (V, S_{\nu_0}(x))_{\Sigma_{\nu_0}^{-1}}$$

exists and is equal to the DOWELSE $\hat{\nu}_{\nu_0}$ which is defined by

$$\hat{\nu}_{\nu_0}(x) = \nu^{(0)}(x) = G_{\nu_0}^{-}(V, S_{\nu_0}(x))_{\Sigma_{\nu_0}^{-1}}; x \in \mathcal{X}.$$

By this solution the projection, by the inner product $(.,.)_{\Sigma_\nu^{-1}}$, of the matrix $S_\nu(x)$ on the subspace $\mathcal{L}(V_1, ..., V_l)$, is defined.
Thus *the likelihood equations for a mixed LRM are*

$$\begin{aligned} G_\nu\nu &= (V, S_\nu(x))_{\Sigma_\nu^{-1}}, \\ \beta_\nu^*(x) &= (F'\Sigma_\nu^{-1}F)^{-1}F'\Sigma_\nu^{-1}x, \end{aligned}$$

where the matrix $S_\nu(x)$ is given by

$$S_\nu(x) = (x - F\beta_\nu^*(x))(x - F\beta_\nu^*(x))'.$$

The solution of the last equations should be found iteratively. Let $\nu^{(0)}(x)$ be some initial value of the parameter ν. Then the $(i + 1)$th iteration $\nu^{(i+1)}(x)$ at the point $x \in \mathcal{X}$ is given by

$$\nu^{(i+1)}(x) = G_{\nu^{(i)}}^{-}(V, S_{\nu^{(i)}}(x))_{\Sigma_{\nu^{(i)}}^{-1}}.$$

The iterations can be stopped if

$$\left\| \nu^{(i+1)}(x) - \nu^{(i)}(x) \right\| < \delta,$$

where δ is a given positive number characterizing the precision of the computation. Thus the MLE $\tilde{\nu}(x)$ is computed as iterated DOWELSEs.

The MLEs of parameter $\theta = (\beta', \nu')'$ of a mixed LRM are $\tilde{\nu}(x)$ which is equal to the value of the last iteration and

$$\tilde{\beta}(x) = \beta^*_{\tilde{\nu}(x)}(x) = (F'\Sigma^{-1}_{\tilde{\nu}(x)}F)^{-1}F'\Sigma^{-1}_{\tilde{\nu}(x)}x; x \in \mathcal{X}.$$

Example 3.4.2. Let us consider a time series

$$X(t) = \beta_1 + \beta_2 t + \varepsilon(t); t = 1, 2, ..., n, \beta \in E^2,$$

where $E[\varepsilon(t)] = 0, D[\varepsilon(t)] = \sigma_1^2$ for $t = 1, 2, ..., n_1$, and $D[\varepsilon(t)] = \sigma_2^2$ for $t = n_1 + 1, n_1 + 2, ..., n_2; n_1 + n_2 = n$. Then we can write

$$X = F\beta + \varepsilon; \beta \in E^2, E[\varepsilon] = 0, Cov(X) = \sigma_1^2 V_1 + \sigma_2^2 V_2, \sigma_1^2, \sigma_2^2 > 0,$$

and where

$$V_1 = \begin{pmatrix} I_{n_1} & 0 \\ 0 & 0 \end{pmatrix} \text{ and } V_2 = \begin{pmatrix} 0 & 0 \\ 0 & I_{n_2} \end{pmatrix}.$$

Ninety independent simulations of the random vector X with given values of parameters β and $\nu = (\sigma_1^2, \sigma_2^2)$ and with $n = 100, n_1 = n_2 = 50$, were used to compute the MLEs $\tilde{\beta}_j$ and $\tilde{\nu}_j; j = 1, 2, ..., 90$, of β and ν. Then the arithmetic means $\bar{\beta}$ and $\bar{\nu}$ of $\tilde{\beta}_j$ and $\tilde{\nu}_j; j = 1, 2, ..., 90$, were computed and also the empirical covariance matrix $\hat{\Sigma}^{\tilde{\nu}}$ of the MLE $\tilde{\nu}$ given by

$$\hat{\Sigma}^{\tilde{\nu}} = \frac{1}{90} \sum_{j=1}^{90} (\tilde{\nu}_j - \bar{\nu})(\tilde{\nu}_j - \bar{\nu})'$$

was computed. This was also done for the DOOLSEs $\hat{\nu}_j; j = 1, 2, ..., 90$, which were used as initial to iterations for computing the MLEs $\tilde{\nu}_j; j = 1, 2, ..., 90$. The results for $\beta_1 = 7.2, \beta_2 = -1.5, \nu_1 = .36$, and $\nu_2 = .64$ are as follows. In the following table one computation of the MLE from one simulation is shown:

Iteration i	$\tilde{\beta}_1^{(i)}$	$\tilde{\beta}_2^{(i)}$	$\tilde{\nu}_1^{(i)}$	$\tilde{\nu}_2^{(i)}$
0	7.298	−1.499	.304	.697
1	7.322	−1.500	.303	.698
2	7.322	−1.500	.303	.698

We can see from this table that the OLSE $\hat{\beta} = \tilde{\beta}^{(0)}$ and the DOOLSE $\hat{\nu} = \tilde{\nu}^{(0)}$ are also good estimators in this simulation.

The arithmetic mean of $\hat{\nu}_j; j = 1, 2, ..., 90$, is

$$\bar{\nu} = (.362, .616)'$$

and the estimated covariance matrix is

$$\hat{\Sigma}^{\hat{\nu}} = \begin{pmatrix} .006 & 0 \\ 0 & .015 \end{pmatrix}.$$

The arithmetic mean of the MLEs $\tilde{\nu}_j; j = 1, 2, ..., 90$, is

$$\bar{\nu} = (.361, .617)'$$

and the estimated covariance matrix of the MLE $\tilde{\nu}$ is the same as that for the DOOLSE $\hat{\nu}$:

$$\hat{\Sigma}^{\tilde{\nu}} = \begin{pmatrix} .006 & 0 \\ 0 & .015 \end{pmatrix}.$$

Similar simulation results were obtained for the same β's and for $\nu_1 = .04$ and $\nu_2 = 1.44$. In this case some estimates $\tilde{\nu}_{j,1}$ were small negative numbers. The arithmetic mean of $\hat{\nu}_j; j = 1, 2, ..., 90$, was

$$\bar{\nu} = (.044, 1.405)'$$

and the estimated covariance matrix

$$\hat{\Sigma}^{\hat{\nu}} = \begin{pmatrix} 0 & 0 \\ 0 & .064 \end{pmatrix}.$$

The arithmetic mean of the MLEs $\tilde{\nu}_j; j = 1, 2, ..., 90$, was

$$\bar{\nu} = (.038, 1.423)'$$

and the estimated covariance matrix of the MLE $\tilde{\nu}$ is practically the same as for the DOOLSE $\hat{\nu}$:

$$\hat{\Sigma}^{\tilde{\nu}} = \begin{pmatrix} 0 & 0 \\ 0 & .068 \end{pmatrix}.$$

Example 3.4.3. Let us again consider an observation X of the time series

$$X(t) = \beta_1 + \beta_2 t + \varepsilon(t); t = 1, 2, ...$$

Then we can write

$$X = F\beta + \varepsilon; \beta \in E^2.$$

We assume, as usual, that $E[\varepsilon] = 0$.

Let us assume now that

$$\varepsilon = \varepsilon^{(1)}U_1 + \varepsilon^{(2)}U_2,$$

where $\varepsilon^{(i)}; i = 1, 2$, are independent $N_1(0, \sigma_i^2)$ random variables and $U_1 = I_n$ and $U_{2,ij} = 1$ if $|i - j| = 1$ and $U_{2,ij} = 0$ elsewhere. Then we get

$$\Sigma \in \Xi = \{\Sigma_\nu : \Sigma_\nu = \sigma_1^2 V_1 + \sigma_2^2 V_2; \nu_i = \sigma_i \in (0, \infty); i = 1, 2\},$$

where $V_1 = I_n$ and $V_2 = U_2 U_2$. From these assumptions we get the equalities $\Sigma_{\nu,11} = \Sigma_{\nu,nn} = \sigma_1^2 + \sigma_2^2, \Sigma_{\nu,ii} = \sigma_1^2 + 2\sigma_2^2$ for $i = 2, 3, ..., n - 1, \Sigma_{\nu,ij} = \sigma_2^2$ for $|i - j| = 2$, and $\Sigma_{\nu,ij} = 0$ elsewhere.

Again 90 independent simulations of the random vector X with given values of parameters β and $\nu = (\sigma_1^2, \sigma_2^2)$ and with $n = 100$ were used to compute the MLEs $\tilde{\beta}_j$ and $\tilde{\nu}_j; j = 1, 2, ..., 90$, of β and ν. Then the arithmetic means $\bar{\beta}$ and $\bar{\nu}$ of $\tilde{\beta}_j$ and $\tilde{\nu}_j; j = 1, 2, ..., 90$, were computed and also the empirical covariance matrix $\hat{\Sigma}^{\tilde{\nu}}$ of the MLE $\tilde{\nu}$, given by

$$\hat{\Sigma}^{\tilde{\nu}} = \frac{1}{90} \sum_{j=1}^{90} (\tilde{\nu}_j - \bar{\nu})(\tilde{\nu}_j - \bar{\nu})',$$

was computed. This was also done for the DOOLSEs $\hat{\nu}_j; j = 1, 2, ..., 90$, which were used as initial to the iterations for computing the MLEs $\tilde{\nu}_j; j = 1, 2, ..., 90$.

The results for $\beta_1 = 7.2, \beta_2 = -1.5, \nu_1 = .64$, and $\nu_2 = .36$ are as follows. In the following table one computation of the MLE from one simulation is shown:

Iteration i	$\tilde{\beta}_1^{(i)}$	$\tilde{\beta}_2^{(i)}$	$\tilde{\nu}_1^{(i)}$	$\tilde{\nu}_2^{(i)}$
0	7.047	−1.499	0.778	0.389
1	7.071	−1.499	0.682	0.446
2	7.070	−1.499	0.664	0.458
3	7.070	−1.499	0.661	0.461
4	7.070	−1.499	0.660	0.462

We can see again from this table that the OLSE $\hat{\beta} = \tilde{\beta}^{(0)}$ and the DOOLSE $\hat{\nu} = \tilde{\nu}^{(0)}$ are also good estimators in this simulation. The arithmetic mean of $\hat{\nu}_j; j = 1, 2, ..., 90$, is

$$\bar{\nu} = (.635, .328)'$$

and the estimated covariance matrix

$$\hat{\Sigma}^{\hat{\nu}} = \begin{pmatrix} .044 & -.019 \\ -.019 & .017 \end{pmatrix}.$$

The arithmetic mean of the MLEs $\tilde{\nu}_j; j = 1, 2, ..., 90$, is

$$\bar{\nu} = (.613, .340)'$$

and the estimated covariance matrix of the MLE $\tilde{\nu}$ is

$$\hat{\Sigma}^{\tilde{\nu}} = \begin{pmatrix} .036 & -.014 \\ -.014 & .015 \end{pmatrix}.$$

Similar simulation results were obtained for the same β and for $\nu_1 = 1.44$ and $\nu_2 = 0.04$. In this case some estimates $\tilde{\nu}_{j,1}$ were small negative numbers. The arithmetic mean of the MLEs $\tilde{\nu}_j; j = 1, 2, ..., 90$, was

$$\bar{\nu} = (1.44, .014)'$$

and the estimated covariance matrix of the MLE $\tilde{\nu}$ was

$$\hat{\Sigma}^{\tilde{\nu}} = \begin{pmatrix} .112 & -.041 \\ -.041 & .024 \end{pmatrix}.$$

Let us now consider the case when $X(.)$ is a Gaussian time series and its observation X is given by the L-NRM

$$X = F\beta + \varepsilon; \beta \in E^k, E[\varepsilon] = 0, Cov(X) = \Sigma \in \Xi = \{\Sigma_\nu; \nu \in \Upsilon\},$$

and let $\Sigma_\nu; \nu \in \Upsilon$, *depend on* ν *nonlinearly.* Then again the MLE $\tilde{\nu}$ for ν is

$$\tilde{\nu}(x) = \arg\max_\nu l_x(\nu); x \in \mathcal{X},$$

where

$$l_x(\nu) = -\frac{1}{2}\ln\det(\Sigma_\nu) - \frac{1}{2}\|x - F\beta_\nu^*(x)\|_{\Sigma_\nu^{-1}}^2$$

and

$$\beta_\nu^*(x) = (F'\Sigma_\nu^{-1}F)^{-1}F'\Sigma_\nu^{-1}x; x \in \mathcal{X}.$$

To find the MLE $\tilde{\nu}$ for ν we have to use again some iterative procedure. The description of commonly used iterative procedures for computing the argument of the maximum of any function, and especially of a loglikelihood function, is given in Harville (1977) and in Azais, Bardin, and Dhorne (1993).

The simplest method is the *gradient method* where $\nu^{(i+1)} - \nu^{(i)}$ is proportional to the gradient $\nabla l_x(\nu^{(i)})$ of $l_x(\nu)$, where

$$\nabla l_x(\nu_0) = \frac{\partial l_x(\nu)}{\partial \nu}\Big|_{\nu=\nu_0}.$$

A refinement of this method is obtained using a second-order approximation of $l_x(\nu)$ by means of the Hessian matrix H, where

$$H_{x,ij}(\nu_0) = \frac{\partial^2 l_x(\nu)}{\partial \nu_i \partial \nu_j} \big|_{\nu=\nu_0}; i, j = 1, 2, ..., l.$$

This method is called the *Newton method*.

Since $H_X(\nu)$ is a random matrix, another method consists of replacing the Hessian matrix by its expectation. This is the *Fisher scoring algorithm (FSA)* method. Given $\nu^{(i)}$ the next iteration $\nu^{(i+1)}$ is given by

$$\nu^{(i+1)}(x) = \nu^{(i)}(x) - (E[H_X(\nu^{(i)})])^- \nabla l_x(\nu^{(i)}); x \in \mathcal{X}.$$

Let $X \sim N_n(F\beta, \Sigma_\nu); \beta \in E^k, \nu \in \Upsilon$, and let

$$V_{\nu_0,i} = \frac{\partial \Sigma_\nu}{\partial \nu_i} \big|_{\nu=\nu_0}; i = 1, 2, ..., l.$$

Then it is known that the FSA for computing the MLE $\tilde{\nu}$ for ν is given by

$$\nu^{(i+1)} = \nu^{(i)} + G^-_{\nu^{(i)}}[(S_{\nu^{(i)}}, V_{\nu^{(i)}})_{\Sigma_{\nu^{(i)}}^{-1}} - (\Sigma_{\nu^{(i)}}, V_{\nu^{(i)}})_{\Sigma_{\nu^{(i)}}^{-1}}],$$

where

$$\begin{aligned} G_{\nu,ij} &= (V_{\nu,i}, V_{\nu,j})_{\Sigma_\nu^{-1}}, \\ S_\nu(x) &= (x - F\beta_\nu^*(x))(x - F\beta_\nu^*(x))' \end{aligned}$$

and

$$(S_\nu, V_\nu)_{\Sigma_\nu^{-1}} \text{ and } (\Sigma_\nu, V_\nu)_{\Sigma_\nu^{-1}}$$

denotes the $l \times 1$ vectors with components $(S_\nu, V_{\nu,i})_{\Sigma_\nu^{-1}}$ and $(\Sigma_\nu, V_{\nu,i})_{\Sigma_\nu^{-1}}$; $i = 1, 2, ..., l$, respectively. The iterations are stopped if

$$\left\| \nu^{(i+1)}(x) - \nu^{(i)}(x) \right\| < \delta,$$

where δ is a given positive number characterizing the precision of the computation.

The MLEs of parameter $\theta = (\beta', \nu')'$ are again $\tilde{\nu}(x)$ which is equal to the value of the last iteration and

$$\tilde{\beta}(x) = \beta_{\tilde{\nu}(x)}^*(x) = (F'\Sigma_{\tilde{\nu}(x)}^{-1}F)^{-1}F'\Sigma_{\tilde{\nu}(x)}^{-1}x; x \in \mathcal{X}.$$

Remarks. 1. In the linear case when $\Sigma_\nu = \sum_{j=1}^{l} \nu_j V_j, ; \nu \in \Upsilon$, the matrices

$V_{\nu,i} = V_i$ do not depend on ν and we get

$$G_{\nu^{(i)}}^-(\Sigma_{\nu^{(i)}}, V_{\nu^{(i)}})_{\Sigma_{\nu^{(i)}}^{-1}} = G_{\nu^{(i)}}^-\left(\sum_{j=1}^l \nu_j^{(i)} V_j, V_{\nu^{(i)}}\right)_{\Sigma_{\nu^{(i)}}^{-1}} = G_{\nu^{(i)}}^- G_{\nu^{(i)}} \nu^{(i)} = \nu^{(i)}$$

and FSA reduces to

$$\nu^{(i+1)} = G_{\nu^{(i)}}^-(V, S_{\nu^{(i)}})_{\Sigma_{\nu^{(i)}}^{-1}}.$$

We see that the iterated DOWELSEs which we have used for computing the MLE $\tilde{\nu}$ in a mixed linear prediction regression model are identical with FSA.

2. It is convenient to use the DOOLSE $\hat{\nu}$ as the initial value $\nu^{(0)}$ of iterations according to FSA, since in this case the MLE $\tilde{\nu}$ is an invariant estimator which does not depend on β for every finite n.

The MLEs $\tilde{\beta}(x)$ and $\tilde{\nu}(x)$ are nonlinear functions of $x \in \mathcal{X}$ computed iteratively and thus it is very difficult to express their statistical properties for a finite length n of the observation X. But the asymptotic properties of the MLE $\tilde{\theta} = (\tilde{\beta}', \tilde{\nu}')'$ are known and in many cases the MLE $\tilde{\theta}$ behaves very well. It is often an asymptotically efficient, consistent, and asymptotically normal estimator of $\theta = (\beta', \nu')'$. These results are given in Sweeting (1980), Mardia and Marshall (1984), Harvey (1994), and others. We shall give the basic results on the MLE $\tilde{\theta}$ which can be found in more detail in Mardia and Marshall (1984).

Let $X(.)$ be a Gaussian time series and let its finite observation X of the length n be an $N_n(F\beta, \Sigma_\nu)$ distributed random vector, where $\beta \in E^k, \nu \in \Upsilon$, and $\Sigma_{\nu,st} = R_\nu(s,t); s,t = 1,2,...,n$. Then the Fisher information matrix $I(\theta); \theta = (\beta', \nu')'$, of X is

$$I(\theta) = \begin{pmatrix} F'\Sigma_\nu^{-1}F & 0 \\ 0 & \frac{1}{2}G_\nu \end{pmatrix},$$

where, as before, $G_{\nu,ij} = (V_{\nu,i}, V_{\nu,j})_{\Sigma_\nu^{-1}}$ and $V_{\nu_0,i} = \partial\Sigma_\nu/\partial\nu_i \mid_{\nu=\nu_0}; i = 1,2,...,l$.

Let

$$V_{\nu_0,ij} = \frac{\partial^2\Sigma_\nu}{\partial\nu_i\partial\nu_j} \mid_{\nu=\nu_0}; i,j = 1,2,...,l,$$

let $\lambda_1 \leq \lambda_2 \leq ... \leq \lambda_n$ be the eigenvalues of Σ_ν, and let those of $V_{\nu,i}$ and $V_{\nu,jj}$ be $\lambda_{i,k}$ and $\lambda_{ij,k}$ for $k = 1,2,...,n$, respectively, with $|\lambda_{i,1}| \leq |\lambda_{i,2}| \leq ... \leq |\lambda_{i,n}|$ and $|\lambda_{ij,1}| \leq |\lambda_{ij,2}| \leq ... \leq |\lambda_{ij,n}|$ for $i,j = 1,2,...,l$.

The proof of the following theorem can be found in Mardia and Marshall (1984).

Theorem 3.4.1. *Suppose, as* $n \to \infty$:
(i) for all $i, j = 1, 2, ..., l$ *the following limits exist:*
$\lim \lambda_n = c < \infty, \lim \lambda_{i,n} = c_i < \infty, \lim \lambda_{ij,n} = c_{ij} < \infty;$
(ii) $\|V_{\nu,i}\|^2 = O(n^{-\frac{1}{2}-\delta})$ *for some* $\delta > 0$ *for* $i = 1, 2, ..., l;$
(iii) for all $i, j = 1, 2, ..., l:$

$$A_{\nu,ij} = \lim \frac{G_{\nu,ij}}{(G_{\nu,ii} G_{\nu,jj})^{\frac{1}{2}}}$$

exists and the matrix A_ν *with components* $A_{\nu,ij}; i, j = 1, 2, ..., l$, *is a non-singular matrix;*
(iv) $\lim (F'F)^{-1} = 0.$
Then the MLE $\tilde{\theta} = (\tilde{\beta}', \tilde{\nu}')'$ *is a weakly consistent, asymptotically normal and efficient estimator of* θ; *that is,* $\tilde{\theta} \sim N_{k+l}(\theta, Cov_\theta(\tilde{\theta}))$, *where*

$$Cov_\theta(\tilde{\theta}) = I^{-1}(\theta) = \begin{pmatrix} (F'\Sigma_\nu^{-1}F)^{-1} & 0 \\ 0 & 2G_\nu^{-1} \end{pmatrix}$$

is the asymptotic covariance matrix of the MLE $\tilde{\theta} = (\tilde{\beta}', \tilde{\nu}')'$.
 If $X(.)$ is covariance stationary with covariance functions $R_\nu(.); \nu \in \Upsilon$, and their derivatives

$$R_{\nu,i}(.) = \frac{\partial R_\nu(.)}{\partial \nu_i} \text{ and } R_{\nu,ij}(.) = \frac{\partial^2 R_\nu(.)}{\partial \nu_i \partial \nu_j}; i, j = 1, 2, ..., l,$$

then we have the following theorem.
 Theorem 3.4.2. *For a covariance stationary Gaussian time series subject to conditions (iii) and (iv) of Theorem 3.4.1 the MLE* $\tilde{\theta} = (\tilde{\beta}', \tilde{\nu}')'$ *is a weakly consistent, asymptotically normal, and efficient estimator of* $\theta = (\beta', \nu')'$ *if*

$$\sum_{t=0}^{\infty} |R_\nu(t)| < \infty, \sum_{t=0}^{\infty} |R_{\nu,i}(t)| < \infty, \text{ and } \sum_{t=0}^{\infty} |R_{\nu,ij}(t)| < \infty,$$

for all $i, j = 1, 2, ..., l$.
 Example 3.4.4. In Example 3.4.1 we derived the MLE of ν for an AR(1) time series

$$X(t) = \rho X(t-1) + \varepsilon(t); t = 1, 2, ...,$$

with $\nu = (\rho, \sigma^2)'$, where $\rho \in (-1, 1)$ and $\sigma^2 \in (0, \infty)$. It is well known that $X(.)$ is stationary with $E[X(t)] = 0$ and

$$R_\nu(t) = \frac{\sigma^2}{1 - \rho^2} \rho^t; t = 0, 1, 2, ...$$

Since

$$\sum_{t=0}^{\infty} \rho^t < \infty, \sum_{t=0}^{\infty} t\rho^t < \infty, \text{ and } \sum_{t=0}^{\infty} t^2\rho^t < \infty \text{ if } \rho \in (-1,1),$$

the conditions

$$\sum_{t=0}^{\infty} |R_\nu(t)| < \infty, \sum_{t=0}^{\infty} |R_{\nu,i}(t)| < \infty, \text{ and } \sum_{t=0}^{\infty} |R_{\nu,ij}(t)| < \infty$$

of Theorem 3.4.2 are fulfilled. Also the condition (iv) of this theorem holds. Let us compute the matrix G_ν.

Using the results of Example 3.4.1 we can write

$$V_{\nu,1} = \frac{\partial \Sigma_\nu}{\partial \rho} = \frac{2\rho}{1 - \rho^2} \Sigma_\nu + \frac{\sigma^2}{1 - \rho^2} V_\rho,$$

where

$$V_\rho = \begin{pmatrix} 0 & 1 & \cdots & (n-2)\rho^{n-3} & (n-1)\rho^{n-2} \\ 1 & 0 & \cdots & (n-3)\rho^{n-4} & (n-2)\rho^{n-3} \\ \cdot & \cdot & \cdots & \cdot & \cdot \\ (n-2)\rho^{n-3} & (n-1)\rho^{n-2} & \cdots & 0 & 1 \\ (n-1)\rho^{n-2} & (n-2)\rho^{n-3} & \cdots & 1 & 0 \end{pmatrix}$$

and

$$V_{\nu,2} = \frac{\partial \Sigma_\nu}{\partial \sigma^2} = \frac{1}{\sigma^2} \Sigma_\nu.$$

Next it is easy to find that

$$\frac{\sigma^2}{1 - \rho^2} V_\rho \Sigma_\nu^{-1} = \frac{1}{1 - \rho^2} B_\rho,$$

where

$$B_\rho = \begin{pmatrix} -\rho & 1 & \rho^2 & \cdots & \rho^{n-2} \\ 1 - \rho^2 & -2\rho & \rho(1-\rho^2) & \cdots & \rho^{n-3}(1-\rho^2) \\ \rho(1-\rho^2) & 1-\rho^2 & \frac{1-\rho^2}{} & \cdots & \rho^{n-4}(1-\rho^2) \\ \cdot & \cdot & \cdots & \cdot \\ \rho^{n-3}(1-\rho^2) & \rho^{n-4}(1-\rho^2) & \rho^{n-6}(1-\rho^2) & \cdots & 1-\rho^2 \\ \rho^{n-2} & \rho^{n-3} & \rho^{n-5} & \cdots & -\rho \end{pmatrix}.$$

Thus we get

$$
\begin{aligned}
G_{\nu,11} &= (V_{\nu,1}, V_{\nu,1})_{\Sigma_\nu^{-1}} = tr(V_{\nu,1}\Sigma_\nu^{-1}V_{\nu,1}\Sigma_\nu^{-1}) \\
&= tr\left(\left(\frac{2\rho}{1-\rho^2}\Sigma_\nu + \frac{\sigma^2}{1-\rho^2}V_\rho\right)\Sigma_\nu^{-1}\left(\frac{2\rho}{1-\rho^2}\Sigma_\nu + \frac{\sigma^2}{1-\rho^2}V_\rho\right)\Sigma_\nu^{-1}\right) \\
&= \left(\frac{2\rho}{1-\rho^2}\right)^2 tr(I_n) + 2\frac{2\rho}{(1-\rho^2)^2}tr(B_\rho) + \frac{1}{(1-\rho^2)^2}tr((B_\rho)^2) \\
&= \frac{4\rho^2 n - 8\rho^2(n-1) + tr((B_\rho)^2)}{(1-\rho^2)^2} \\
&= \frac{-4\rho^2(n-2) + tr((B_\rho)^2)}{(1-\rho^2)^2}.
\end{aligned}
$$

Next we have

$$
tr((B_\rho)^2) = \sum_{i=1}^{n}\sum_{k=1}^{n}B_{\rho,ik}B_{\rho,ki}
$$

and, using the equalities,

$$
\begin{aligned}
tr((B_\rho)^2) &= \sum_{k=1}^{n}B_{\rho,1k}B_{\rho,k1} + \sum_{k=1}^{n}B_{\rho,nk}B_{\rho,kn} + \sum_{i=2}^{n-1}\sum_{k=1}^{n}B_{\rho,ik}B_{\rho,ki} \\
&= \sum_{k=1}^{n}B_{\rho,1k}B_{\rho,k1} + \sum_{k=1}^{n}B_{\rho,nk}B_{\rho,kn} \\
&\quad + \sum_{i=2}^{n-1}B_{\rho,i1}B_{\rho,1i} + \sum_{i=2}^{n-1}B_{\rho,in}B_{\rho,ni} + \sum_{i=2}^{n-1}\sum_{k=2}^{n-1}B_{\rho,ik}^2 \\
&= B_{\rho,11}^2 + 2B_{\rho,1n}B_{\rho,n1} + B_{\rho,nn}^2 + 2\sum_{k=2}^{n-1}B_{\rho,1k}B_{\rho,k1} \\
&\quad + 2\sum_{k=2}^{n-1}B_{\rho,nk}B_{\rho,kn} + \sum_{i=2}^{n-1}\sum_{k=2}^{n-1}B_{\rho,ik}^2,
\end{aligned}
$$

since $B_{\rho,ik} = B_{\rho,ki}$ for $i, k = 2, ..., n-1$, we get

$$
\begin{aligned}
tr((B_\rho)^2) &= 2\rho^2 + 2\rho^{2(n-2)} + 4(1-\rho^2)^2\sum_{k=0}^{n-3}(\rho^2)^k \\
&\quad + (n-2)4\rho^2 + 2(1-\rho^2)^2\sum_{k=1}^{n-3}(n-2-k)(\rho^2)^{k-1}.
\end{aligned}
$$

After a simple algebra, using the equalities

$$\sum_{k=0}^{n-3}(\rho^2)^k = \frac{1-\rho^{2(n-2)}}{1-\rho^2}\sum_{k=1}^{n-3}(n-2-k)(\rho^2)^{k-1}$$

$$= \frac{\rho^{2(n-1)}+n(1-\rho^2)+2\rho^2-3}{(1-\rho^2)^2},$$

we get

$$tr((B_\rho)^2) = 2\rho^2 + 2\rho^{2(n-2)} + 4(1-\rho^2)(1-\rho^{2(n-2)})$$
$$+(n-2)4\rho^2 + 2\rho^{2(n-1)} + 2n(1-\rho^2) + 4\rho^2 - 6$$

and, after some algebra,

$$tr((B_\rho)^2) = 2n(1+\rho^2) - 2\rho^{2(n-2)} + 6\rho^{2(n-1)} - 6\rho^2 - 2.$$

Thus

$$G_{\nu,11} = \frac{-4\rho^2(n-2)+2n(1+\rho^2)-2\rho^{2(n-2)}+6\rho^{2(n-1)}-6\rho^2-2}{(1-\rho^2)^2}$$

$$= \frac{2(n-1)(1-\rho^2)-2\rho^{2(n-2)}+6\rho^{2(n-1)}}{(1-\rho^2)^2}.$$

Next we have

$$G_{\nu,12} = (V_{\nu,1}, V_{\nu,2})_{\Sigma_\nu^{-1}} = tr(V_{\nu,1}\Sigma_\nu^{-1}V_{\nu,2}\Sigma_\nu^{-1})$$

$$= tr\left(\left(\frac{2\rho}{1-\rho^2}\Sigma_\nu + \frac{\sigma^2}{1-\rho^2}V_\rho\right)\Sigma_\nu^{-1}\frac{1}{\sigma^2}\Sigma_\nu.\Sigma_\nu^{-1}\right)$$

$$= \frac{2\rho}{\sigma^2(1-\rho^2)}n + \frac{1}{\sigma^2(1-\rho^2)}tr(B_\rho)$$

$$= \frac{2\rho n - 2\rho(n-1)}{\sigma^2(1-\rho^2)} = \frac{2\rho}{\sigma^2(1-\rho^2)}$$

and

$$G_{\nu,22} = (V_{\nu,2}, V_{\nu,2})_{\Sigma_\nu^{-1}} = tr(V_{\nu,2}\Sigma_\nu^{-1}V_{\nu,2}\Sigma_\nu^{-1}) = \frac{1}{\sigma^4}tr(I_n) = \frac{1}{\sigma^4}n.$$

Thus

$$A_{\nu,12} = \lim\frac{G_{\nu,12}}{(G_{\nu,11}G_{\nu,22})^{\frac{1}{2}}} = 0,$$

$$A_{\nu,11} = A_{\nu,22} = 1,$$

and the matrices $A_\nu; \nu \in \Upsilon$ are nonsingular.

Thus the MLE $\tilde{\nu} = (\tilde{\rho}, \tilde{\sigma}^2)'$ is a weakly consistent, asymptotically normal, and efficient estimator of $\nu = (\rho, \sigma^2)'$, that is, $\tilde{\nu} \sim N_2(\nu, 2G_\nu^{-1})$, where the elements of the matrix G_ν^{-1} are:

$$
\begin{aligned}
(G_\nu^{-1})_{11} &= 1/(1-\rho^2)^2 \left(2(n-1)(1-\rho^2) - 2\rho^{2(n-2)} + 6\rho^{2(n-1)}\right), \\
(G_\nu^{-1})_{12} &= 2\rho(1-\rho^2)/\sigma^2, \\
(G_\nu^{-1})_{22} &= n(1-\rho^2)^2/\sigma^4
\end{aligned}
$$

Thus, since $Cov_\nu(\tilde{\nu}) = 2G_\nu^{-1}$, we can write the asymptotic covariance matrix as

$$
Cov_\nu(\tilde{\nu}) \approx \frac{1}{n}\begin{pmatrix} 1-\rho^2 & 0 \\ 0 & 2\sigma^4 \end{pmatrix}
$$

from which *the asymptotic variances and covariance of the MLE* $\tilde{\nu} = (\tilde{\rho}, \tilde{\sigma}^2)'$ are

$$
\begin{aligned}
D_\nu[\tilde{\rho}] &\approx \frac{1}{n}(1-\rho^2), \\
D_\nu[\tilde{\sigma}^2] &\approx \frac{2\sigma^4}{n},
\end{aligned}
$$

and

$$
Cov_\nu[\tilde{\rho}, \tilde{\sigma}^2] \approx 0.
$$

Example 3.4.5. Let us consider a zero mean, covariance stationary time series $X(.)$ with covariance functions

$$
R_\nu(t) = \sigma^2 e^{-\alpha t}; t = 0, 1, 2, ...,
$$

where $\nu = (\sigma^2, \alpha)' \in \Upsilon = (0, \infty) \times (0, \infty)$.

Since

$$
\sum_{t=0}^{\infty} e^{-\alpha t} < \infty \text{ and } \sum_{t=0}^{\infty} te^{-\alpha t} < \infty \text{ if } \alpha \in (0, \infty),
$$

the conditions

$$
\sum_{t=0}^{\infty} |R_\nu(t)| < \infty, \sum_{t=0}^{\infty} |R_{\nu,i}(t)| < \infty, \text{ and } \sum_{t=0}^{\infty} |R_{\nu,ij}(t)| < \infty
$$

of Theorem 3.4.2 are fulfilled. Also condition (*iv*) of this theorem holds.

Let us compute the matrix G_ν. We have

$$\Sigma_\nu = \sigma^2 \begin{pmatrix} 1 & e^{-\alpha} & \cdots & e^{-(n-2)\alpha} & e^{-(n-1)\alpha} \\ e^{-\alpha} & 1 & \cdots & e^{-(n-3)\alpha} & e^{-(n-2)\alpha} \\ \cdot & & \cdots & & \cdot \\ e^{-(n-2)\alpha} & e^{-(n-3)\alpha} & \cdots & 1 & e^{-\alpha} \\ e^{-(n-1)\alpha} & e^{-(n-2)\alpha} & \cdots & e^{-\alpha} & 1 \end{pmatrix}$$

and

$$\Sigma_\nu^{-1} = \frac{1}{\sigma^2(1 - e^{-2\alpha})} \begin{pmatrix} 1 & -e^{-\alpha} & \cdots & 0 & 0 \\ -e^{-\alpha} & 1 + e^{-2\alpha} & \cdots & 0 & 0 \\ \cdot & \cdot & \cdots & \cdot & \cdot \\ 0 & 0 & \cdots & 1 + e^{-\alpha 2} & -e^{-\alpha} \\ 0 & 0 & \cdots & -e^{-\alpha} & 1 \end{pmatrix}.$$

Thus

$$V_{\nu,1} = \frac{\partial \Sigma_\nu}{\partial \sigma^2} = \frac{1}{\sigma^2} \Sigma_\nu$$

and

$$G_{\nu,11} = (V_{\nu,1}, V_{\nu,1})_{\Sigma_\nu^{-1}} = tr(V_{\nu,1}\Sigma_\nu^{-1}V_{\nu,1}\Sigma_\nu^{-1}) = \frac{1}{\sigma^4}tr(I_n) = \frac{n}{\sigma^4}.$$

After some computation we get also the expression

$$tr(\Sigma_\nu^{-1}V_{\nu,2}) = \frac{2(n-1)e^{-2\alpha}}{1 - e^{-2\alpha}}.$$

Using this result we have

$$\begin{aligned} G_{\nu,12} &= (V_{\nu,1}, V_{\nu,2})_{\Sigma_\nu^{-1}} = \frac{1}{\sigma^2}tr(\Sigma_\nu^{-1}V_{\nu,2}) \\ &= \frac{2(n-1)e^{-2\alpha}}{\sigma^2(1 - e^{-2\alpha})}. \end{aligned}$$

and

$$G_{\nu,22} = (V_{\nu,2}, V_{\nu,2})_{\Sigma_\nu^{-1}} = tr((V_{\nu,2}\Sigma_\nu^{-1})^2).$$

To compute $tr((V_{\nu,2}\Sigma_\nu^{-1})^2)$ we can use the same approach as that for computing $tr((B_\rho)^2)$ in the preceding example, because $(V_{\nu,2}\Sigma_\nu^{-1})^2)$ and $(B_\rho)^2$ are of the same shape.

Thus we get

$$(1 - e^{-2\alpha})^2 tr((V_{\nu,2}\Sigma_\nu^{-1})^2) = 2e^{-4\alpha} + 2e^{-2(n-1)\alpha}$$

$$+2(1 - e^{-2\alpha}) \sum_{k=1}^{n-2} (e^{-2\alpha})^k$$

$$-2(1 - e^{-2\alpha}) \sum_{k=2}^{n-1} (e^{-2\alpha})^k + 4(n-2)e^{-4\alpha}$$

$$+2(1 - e^{-2\alpha})^2 \sum_{k=1}^{n-3} (n - k - 2)(e^{-2\alpha})^k.$$

Using the expressions

$$\sum_{k=1}^{n-2} (e^{-2\alpha})^k = \frac{1 - e^{-2\alpha(n-1)}}{1 - e^{-2\alpha}} \quad \sum_{k=2}^{n-1} (e^{-2\alpha})^k = \frac{e^{-4\alpha} - e^{-2\alpha n}}{1 - e^{-2\alpha}}$$

and

$$\sum_{k=1}^{n-3} (n - k - 2)(e^{-2\alpha})^k = \frac{e^{-2\alpha(n-1)} - (n-2)e^{-4\alpha} + (n-3)e^{-2\alpha}}{(1 - e^{-2\alpha})^2}$$

we get

$$G_{\nu,22} = \frac{1}{(1 - e^{-2\alpha})^2} [2e^{-4\alpha} + 2e^{-2\alpha(n-1)} + 2 - 2e^{-2\alpha(n-1)}$$

$$-2e^{-4\alpha} + 2e^{-2\alpha n} + 4ne^{-4\alpha} - 8e^{-4\alpha}$$

$$+2e^{-2\alpha(n-1)} - 2ne^{-4\alpha} + 4e^{-4\alpha} + 2ne^{-2\alpha} - 6e^{-2\alpha}]$$

and, after some algebra,

$$G_{\nu,22} = \frac{2ne^{-4\alpha}(1 + e^{2\alpha}) + 2e^{-2\alpha n}(1 + e^{2\alpha}) - 2e^{-4\alpha}(2 - 3e^{2\alpha}) + 2}{(1 - e^{-2\alpha})^2}.$$

Thus

$$A_{\nu,12} = \lim \frac{G_{\nu,12}}{(G_{\nu,11}G_{\nu,22})^{\frac{1}{2}}} = \left(\frac{2}{1 + e^{2\alpha}}\right)^{\frac{1}{2}} < 1 \text{ if } \alpha > 0,$$

$$A_{\nu,11} = A_{\nu,22} = 1$$

and we see that the matrices $A_\nu; \nu \in \Upsilon$, are nonsingular and thus condition
(iii) of Theorem 3.4.2 is fulfilled. Thus the MLE $\tilde{\nu} = (\tilde{\sigma}^2, \tilde{\alpha})'$ is a weakly

consistent, asymptotically normal, and efficient estimator of $\nu = (\sigma^2, \alpha)'$, that is $\tilde{\nu} \sim N_2(\nu, 2G_\nu^{-1})$. It can be seen, after some computation, that *the asymptotic covariance matrix $Cov_\nu(\tilde{\nu})$ of the MLE $\tilde{\nu} = (\tilde{\sigma}^2, \tilde{\alpha})'$ is*

$$Cov_\nu(\tilde{\nu}) = 2G_\nu^{-1} \approx \frac{1}{n}\begin{pmatrix} 2\sigma^4\left(1 + e^{-2\alpha}\right)/\left(1 - e^{-2\alpha}\right) & -2\sigma^2 \\ -2\sigma^2 & \left(1 - e^{-2\alpha}\right)/e^{-2\alpha} \end{pmatrix}.$$

For $\sigma^2 = 1, \alpha = 2$, and $n = 10$ we get

$$Cov_\nu(\tilde{\nu}) \approx 2G_\nu^{-1} = \begin{pmatrix} .207 & -.2 \\ -.2 & 5.3598 \end{pmatrix}.$$

The asymptotic variance

$$D_\nu[\tilde{\alpha}] = D_\alpha[\tilde{\alpha}] = \frac{1}{n}(e^{2\alpha} - 1)$$

of the MLE $\tilde{\alpha}$ is an increasing function of α, and can be large for large values of α and for relatively large values of n. For example, if $\alpha = 3$, then

$$D_\alpha[\tilde{\alpha}] = \frac{1}{n}402.43$$

and for $\alpha = 4$ we get $D_\alpha[\tilde{\alpha}] = 1/n\, 2980$ and thus n has to be very large to make this variance small. It should be remarked that in the case when α is very large it is reasonable to consider the time series $X(.)$ as a white noise.

Example 3.4.6. Let us consider a time series with the covariance functions

$$R_\nu(t) = \sigma^2 e^{-\alpha t} \cos \beta t,$$

where $\nu = (\sigma^2, \alpha, \beta)' \in \Upsilon = (0, \infty) \times (0, \infty) \times [0, \pi]$. Then it is easy to see that $V_{\nu,1}\Sigma_\nu^{-1} = 1/\sigma^2 I_n$ and $V_{\nu,2}\Sigma_\nu^{-1}, V_{\nu,3}\Sigma_\nu^{-1}$ do not depend on σ^2. Since the exact expression for Σ_ν^{-1} is not known, we shall give some results for given values of parameter ν.

Let $\nu = \nu_0 = (\sigma_0^2, \alpha_0, \beta_0)' \in \Upsilon$, where $\sigma_0^2 = 1, \alpha_0 = 2, \beta_0 = \pi/4$ and $n = 10$. Then

$$\Sigma_{\nu_0} = \begin{pmatrix} 1 & .01 & 0 & \cdots & 0 & 0 & 0 \\ .01 & 1 & .01 & \cdots & 0 & 0 & 0 \\ 0 & .01 & 1 & \cdots & 0 & 0 & 0 \\ \cdot & \cdot & \cdot & \cdots & \vdots & \cdot & \cdot \\ 0 & 0 & 0 & \cdots & 1 & .01 & 0 \\ 0 & 0 & 0 & \cdots & .01 & 1 & .01 \\ 0 & 0 & 0 & \cdots & 0 & .01 & 1 \end{pmatrix},$$

$$\Sigma_{\nu_0}^{-1} = \begin{pmatrix} 1 & -.01 & 0 & . & 0 & 0 & 0 \\ -.01 & 1 & -.01 & . & 0 & 0 & 0 \\ 0 & -.01 & 1 & . & 0 & 0 & 0 \\ . & . & . & \cdots & -.01 & 0 & 0 \\ 0 & 0 & 0 & . & 1 & -.01 & 0 \\ 0 & 0 & 0 & . & -.01 & 1 & -.01 \\ 0 & 0 & 0 & . & 0 & -.01 & 1 \end{pmatrix},$$

$$V_{\nu_0,2} = \begin{pmatrix} 0 & -.1 & 0 & \cdots & 0 & 0 & 0 \\ -.1 & 0 & -.1 & \cdots & 0 & 0 & 0 \\ 0 & -.1 & 0 & \cdots & 0 & 0 & 0 \\ . & . & . & \cdots & . & . & . \\ 0 & 0 & 0 & \cdots & 0 & -.1 & 0 \\ 0 & 0 & 0 & \cdots & -.1 & 0 & -.1 \\ 0 & 0 & 0 & \cdots & 0 & -.1 & 0 \end{pmatrix},$$

and

$$V_{\nu_0,3} = \begin{pmatrix} 0 & -.1 & -.03 & \cdots & 0 & 0 & 0 \\ -.1 & 0 & -.1 & \cdots & 0 & 0 & 0 \\ -.03 & -.1 & 0 & \cdots & 0 & 0 & 0 \\ . & . & . & \cdots & . & . & . \\ 0 & 0 & 0 & \cdots & 0 & -.1 & -.03 \\ 0 & 0 & 0 & \cdots & -.1 & 0 & -.1 \\ 0 & 0 & 0 & \cdots & -.03 & -.1 & 0 \end{pmatrix}.$$

After some algebra, using these matrices, we get, writing ν instead of ν_0, for simplicity of notation,

$$\begin{aligned} G_{\nu,12} &= tr(\Sigma_\nu^{-1} V_{\nu,2}) = 0.018, \quad G_{\nu,22} = tr((\Sigma_\nu^{-1} V_{\nu,2})^2) = 0.180, \\ G_{\nu,13} &= tr(\Sigma_\nu^{-1} V_{\nu,3}) = 0.018, \quad G_{\nu,33} = tr((\Sigma_\nu^{-1} V_{\nu,3})^2) = 0.193, \\ G_{\nu,23} &= tr(\Sigma_\nu^{-1} V_{\nu,2} \Sigma_\nu^{-1} V_{\nu,3}) = 0.036. \end{aligned}$$

Thus, since $G_{\nu,11} = n/\sigma^4$,

$$G_{\nu_0} = \begin{pmatrix} 10 & 0.01 & 0.01 \\ 0.01 & 0.18 & 0.03 \\ 0.01 & 0.03 & 0.19 \end{pmatrix}$$

and

$$Cov_{\nu_0}(\tilde{\nu}) \approx 2G_{\nu_0}^{-1} = \begin{pmatrix} 0.20 & -0.01 & -0.01 \\ -0.01 & 11.54 & -2.15 \\ -0.01 & -2.15 & 10.76 \end{pmatrix}.$$

For $\sigma^2 \neq 1$ we obtained the following results. If $\nu = (2, 2, \pi/4)'$ and $n = 10$, then

$$G_\nu = \begin{pmatrix} 2.5 & .09 & .09 \\ .09 & 0.18 & 0.03 \\ .09 & 0.03 & 0.19 \end{pmatrix}$$

and

$$Cov_\nu(\tilde{\nu}) \approx 2G_\nu^{-1} = \begin{pmatrix} .80 & -.03 & -.03 \\ -.03 & 11.54 & -2.15 \\ -.03 & -2.15 & 10.76 \end{pmatrix}.$$

If $\nu = (1/2, 2, \pi/4)'$ and $n = 10$, then

$$G_\nu = \begin{pmatrix} 40 & 0.03 & 0.03 \\ 0.03 & 0.18 & 0.03 \\ 0.03 & 0.03 & 0.19 \end{pmatrix}$$

and

$$Cov_\nu(\tilde{\nu}) \approx 2G_\nu^{-1} = \begin{pmatrix} 0.05 & 0 & 0 \\ 0 & 11.54 & -2.15 \\ 0 & -2.15 & 10.76 \end{pmatrix}.$$

It can be seen from these computations that only the variance $D_\nu[\tilde{\sigma}^2]$ of an estimator $\tilde{\sigma}^2$ of σ^2 depends on σ^2 and the variances of the estimators $\tilde{\alpha}$ and $\tilde{\beta}$ are independent of σ^2. In this part of our example the value $\alpha_0 = 2$ of the parameter α was relatively large and as a consequence we have obtained large values of variances of the estimators $\tilde{\alpha}$ and $\tilde{\beta}$. It can be expected, similar to Example 3.4.5, that for smaller values of α these variances will also be smaller.

Let us consider the case when the parameter ν is equal to $\nu_1 = (1, 0.1, \pi/4)'$ and $n = 10$. Then

$$\Sigma_{\nu_1} = \begin{pmatrix} 1 & .64 & 0 & \cdots & .35 & .45 & .29 \\ .64 & 1 & .64 & \cdots & 0 & .35 & .45 \\ 0 & .64 & 1 & \cdots & -.43 & 0 & .35 \\ . & . & . & \cdots & . & . & . \\ .35 & 0 & -.43 & \cdots & 1 & .64 & 0 \\ .45 & .35 & 0 & \cdots & .64 & 1 & .64 \\ .29 & .45 & .35 & \cdots & 0 & .64 & 1 \end{pmatrix},$$

$$\Sigma_{\nu_1}^{-1} = \begin{pmatrix} 3.5 & -3.1 & 1.7 & \cdots & .1 & 0 & 0 \\ -3.1 & 6.3 & -4.6 & \cdots & -.1 & .1 & 0 \\ 1.7 & -4.6 & 7.1 & \cdots & .1 & -.1 & .1 \\ . & . & . & \cdots & . & . & . \\ .1 & -.1 & .1 & \cdots & 7.1 & -4.6 & 1.7 \\ 0 & .1 & -.1 & \cdots & -4.6 & 6.3 & -3.1 \\ 0 & 0 & .1 & \cdots & 1.7 & -3.1 & 3.5 \end{pmatrix}.$$

The elements of matrices $V_{\nu_1,2}$ and $V_{\nu_1,3}$ are given by

$$V_{\nu_1,2,st} = R_{\nu_1,2}(|s - t|) \text{ and } V_{\nu_1,3,st} = R_{\nu_1,3}(|s - t|); s,t = 1, 2, ..., 10.$$

The desired values are in the following table:

t	0	1	2	3	4	5	6	7	8	9
$R_{\nu_1,2}(t)$	0	-0.6	0	1.6	2.7	2.1	0	-2.6	-3.6	-2.6
$R_{\nu_1,3}(t)$	0	-0.6	-1.6	-1.6	0	2.1	3.3	2.6	0	-2.6

After some algebra, using these values, we get, writing only ν instead of ν_1 for simplicity of notation,

$$
\begin{aligned}
G_{\nu,12} &= tr(\Sigma_\nu^{-1} V_{\nu,2}) = 66.304, \\
G_{\nu,22} &= tr((\Sigma_\nu^{-1} V_{\nu,2})^2) = 673.42, \\
G_{\nu,13} &= tr(\Sigma_\nu^{-1} V_{\nu,3}) = 4.4507, \\
G_{\nu,33} &= tr((\Sigma_\nu^{-1} V_{\nu,3})^2) = 121.61,
\end{aligned}
$$

and

$$G_{\nu,23} = tr(\Sigma_\nu^{-1} V_{\nu,2} \Sigma_\nu^{-1} V_{\nu,3}) = 40.915.$$

Thus, since $G_{\nu,11} = n/\sigma^4$,

$$
G_{\nu_1} = \begin{pmatrix}
10 & 66.30 & 4.45 \\
66.30 & 673.42 & 40.91 \\
4.45 & 40.91 & 121.61
\end{pmatrix}
$$

and

$$
Cov_{\nu_1}(\tilde{\nu}) \approx 2G_{\nu_1}^{-1} = \begin{pmatrix}
.57 & -.05 & 0 \\
-.05 & 0 & 0 \\
0 & 0 & 0.01
\end{pmatrix}.
$$

We can see that in this case when α is small, $\alpha = 0.1$, we also get small values of the approximate variances of the components of the MLE $\tilde{\nu}$.

There is also a dependence of these variances on σ^2. For other values of σ^2 as 1 we obtained the following results.

If $\nu = (2, 0.1, \pi/4)'$ and $n = 10$, then

$$
G_\nu = \begin{pmatrix}
2.50 & 33.15 & 2.22 \\
33.15 & 673.42 & 40.91 \\
2.22 & 40.91 & 121.61
\end{pmatrix}
$$

and

$$Cov_\nu(\tilde{\nu}) \approx 2G_\nu^{-1} = \begin{pmatrix} 2.30 & -.11 & 0 \\ -.11 & .01 & 0 \\ 0 & 0 & .01 \end{pmatrix}.$$

If $\nu = (1/2, 0.1, \pi/4)'$ and $n = 10$, then

$$G_\nu = \begin{pmatrix} 40 & 132.60 & 8.90 \\ 132.60 & 673.42 & 40.91 \\ 8.90 & 40.91 & 121.61 \end{pmatrix}$$

and

$$Cov_\nu(\tilde{\nu}) \approx 2G_\nu^{-1} = \begin{pmatrix} .14 & -.02 & 0 \\ -.02 & .01 & 0 \\ 0 & 0 & .01 \end{pmatrix}.$$

It can again be seen from these computations that only the variance $D_\nu[\tilde{\sigma}^2]$ of an estimator $\tilde{\sigma}^2$ depends on σ^2 and the variances of the estimators $\tilde{\alpha}$ and $\tilde{\beta}$ are independent of σ^2.

In the case when the time series $X(.)$ are Gaussian and follow NRMs

$$X(t) = m_\gamma(t) + \varepsilon(t); t \in T, \gamma \in \Upsilon,$$

finite observations X of $X(.)$ are given by

$$X = m_\gamma + \varepsilon; \gamma \in \Gamma \subset E^k, Cov(X) \in \Xi = \left\{ \Sigma_\nu : \Sigma_\nu = \sum_{j=1}^l \nu_j V_j; \nu \in \Upsilon \right\}.$$

This model can also be called *nonlinear regression with variance compo-nents*. In this model the unknown parameter, which should be estimated from X, is $\theta = (\gamma', \nu')' \in \Theta = \Gamma \times \Upsilon$. Under the assumption that the random vector X has n-dimensional normal distributions we can again use the method of maximum likelihood to estimate the unknown parameter θ. The likelihood function which should be maximized is

$$L_x(\theta) = L_x(\gamma, \nu) = \frac{1}{(2\pi)^{n/2} (\det \Sigma_\nu)^{1/2}} \exp\{-\frac{1}{2}(x - m_\gamma)'\Sigma_\nu^{-1}(x - m_\gamma)\}.$$

The MLE $\tilde{\theta}$ of θ can be computed iteratively by using the FSA. This is in more detail described in Gumpertz and Pantula (1992). Let

$$F_{\gamma_0} = \frac{\partial m_\gamma}{\partial \gamma} \Big|_{\gamma=\gamma_0}$$

and

$$S_{\gamma_0}(X) = (X - m_{\gamma_0})(X - m_{\gamma_0})'.$$

Then the expressions for FSA estimates at the $(i + 1)$ iteration are

$$\gamma^{(i+1)}(x) = \gamma^{(i)}(x) + (F'_{\gamma^{(i)}} \Sigma^{-1}_{\nu^{(i)}} F_{\gamma^{(i)}})^{-1} F'_{\gamma^{(i)}} \Sigma^{-1}_{\nu^{(i)}} (x - m_{\gamma^{(i)}})$$

and

$$\nu^{(i+1)}(x) = G^{-1}_{\nu^{(i)}} (V, S_{\gamma^{(i)}})_{\Sigma^{-1}_{\nu^{(i)}}} \; ; i = 0, 1, ...,$$

where $G^{-1}_{\nu^{(i)}}$ is the inverse of the Gramm matrix $G_{\gamma^{(i)}}$ defined, as usual, by

$$G_{\gamma^{(i)}, ij} = tr(V_i \Sigma^{-1}_{\nu^{(i)}} V_j \Sigma^{-1}_{\nu^{(i)}}) = (V_i, V_j)_{\Sigma^{-1}_{\nu^{(i)}}} \; ; i, j = 1, 2, ..., l,$$

and $(V, S_{\gamma^{(i)}})_{\Sigma^{-1}_{\nu^{(i)}}}$ denotes the $l \times 1$ random vector with components

$$(V_j, S_{\gamma^{(i)}})_{\Sigma^{-1}_{\nu^{(i)}}} \; ; j = 1, 2, ..., l.$$

The initial values $\gamma^{(0)}$ and $\nu^{(0)}$ should be determined. The iterations are stopped if

$$\left\| \theta^{(i+1)}(x) - \theta^{(i)}(x) \right\| < \delta,$$

where δ is a given positive number characterizing the precision of the computation.

In Gumpertz and Pantulla (1992) sufficient conditions are given under which the MLE $\tilde{\theta} = (\tilde{\gamma}', \tilde{\nu}')'$ is strongly consistent and asymptotically normal. In this case the asymptotic covariance matrix of $\tilde{\theta}$ is

$$Cov_\theta(\tilde{\theta}) \approx \begin{pmatrix} (F'_\gamma \Sigma^{-1}_\nu F_\gamma)^{-1} & 0 \\ 0 & 2G^{-1}_\nu \end{pmatrix}.$$

Finally, in the case when the time series $X(.)$ are Gaussian and follow NRMs

$$X(t) = m_\gamma(t) + \varepsilon(t); t \in T, \gamma \in \Upsilon,$$

and in which the finite observations X of $X(.)$ are given by

$$X = m_\gamma + \varepsilon; \gamma \in \Gamma \subset E^k, Cov(X) \in \Xi = \{\Sigma_\nu; \nu \in \Upsilon\},$$

where the dependence of Σ_ν on ν is a nonlinear one, the MLE $\tilde{\theta} = (\tilde{\gamma}', \tilde{\nu}')'$ of $\theta = (\gamma', \nu')' \in \Theta = \Gamma \times \Upsilon$ should again be computed iteratively.

The FSA iterations are given by, using the notation as before,

$$\gamma^{(i+1)}(x) = \gamma^{(i)}(x) + (F'_{\gamma^{(i)}} \Sigma^{-1}_{\nu^{(i)}} F_{\gamma^{(i)}})^{-1} F'_{\gamma^{(i)}} \Sigma^{-1}_{\nu^{(i)}} (x - m_{\gamma^{(i)}})$$

and

$$\nu^{(i+1)} = \nu^{(i)} + G^{-}_{\nu^{(i)}} [(S_{\gamma^{(i)}}, V_{\nu^{(i)}})_{\Sigma^{-1}_{\nu^{(i)}}} - (\Sigma_{\nu^{(i)}}, V_{\nu^{(i)}})_{\Sigma^{-1}_{\nu^{(i)}}}].$$

The iterations are stopped again, if

$$\left\| \theta^{(i+1)}(x) - \theta^{(i)}(x) \right\| < \delta,$$

where δ is a given positive number characterizing the precision of the computation. It can be expected that the conditions given in Mardia and Marshall (1984) and in Gumpertz and Pantulla (1992) are sufficient for strong consistency and asymptotic normality of the MLE $\tilde{\theta}$ of θ.

4

Predictions of Time Series

4.1 Introduction

The problem of the prediction of time series belongs to the most important problems of the statistical inference of time series. There are many approaches to these problems, possibly the best known is that based on the Box-Jenkins methodology of modeling time series by using ARMA and ARIMA models. another approach is based on modeling time series by regression models. It can be said that the Box-Jenkins methodology is the most popular and there exists a lot of literature on different levels dealing with this approach. We refer to Box and Jenkins (1976), Brockwell and Davis (1987), (1996), and many others. The approach based on regression models, known as *kriging* in engineering literature, mainly in geostatistics, can be found in David (1977), Journel and Hüijbregts (1978), Ripley (1981), and in Christensen (1987), (1991).

The problem of the prediction of time series can generally be formulated as follows. Let $X = (X(1), ..., X(n))'$ be an observation of a time series $X(.)$ and let U be some random variable. Usually $U = X(n + d)$ for some positive integer d and several times we set $d = 1$. The problem is to find an approximation $\tilde{U}(X)$ of the random variable U based on X in such a way that this approximation is a better one in some sense. If $U = X(n + d)$ we call this approximation $\tilde{U}(X)$ a *(d-step) predictor* of $X(n + d)$ and we use the notation $\tilde{X}(n + d)$. The quality of any predictor $\tilde{U}(X)$ is measured by its MSE defined by

$$MSE[\tilde{U}(X)] = E[\tilde{U}(X) - U]^2.$$

The best predictor $U^*(X)$ of U minimizes the MSEs of all predictors $\tilde{U}(X)$ of U:

$$U^*(X) = \arg\min_{\tilde{U}} E[\tilde{U}(X) - U]^2.$$

The best predictor $U^*(X)$ exists for all U and is equal to the conditional mean value $E[U \mid X]$. The problem is that an explicit expression for $E[U \mid X]$ exists only for a few distributions of the random vector $(X', U)'$, and in many cases we get that $E[U \mid X]$ is a nonlinear function of X and thus it is difficult to study its statistical properties. One of the exceptions is the case when the random vector $(X', U)'$ has $(n + 1)-$ *dimensional normal distribution.* In this case

$$U^*(X) = E[U \mid X] = E[U] + r'\Sigma^{-1}(X - E[X]),$$

where $\Sigma = Cov(X)$ and $r = Cov(X, U)$ is an $n \times 1$ vector of covariances between X and U with components $r_t = Cov(X(t); U); t = 1, 2, ..., n$. Thus in the case of normality we see that the best predictor $U^*(X)$ of U is a linear function of X and such predictors are called *linear predictors*. $U^*(X)$ is an *unbiased predictor of U*, since

$$E[U^*(X)] = E[U].$$

Thus in the case of normality the best predictor $U^*(X)$ of U is equal to the *linear* unbiased predictor $E[U \mid X]$.

The MSE of $U^*(X)$ is

$$
\begin{aligned}
MSE[U^*(X)] &= E[U^* - U]^2 \\
&= D[U] - 2Cov(r'\Sigma^{-1}X; U) + D[r'\Sigma^{-1}X] \\
&= D[U] - 2r'\Sigma^{-1}Cov(X; U) + r'\Sigma^{-1}\Sigma\Sigma^{-1}r \\
&= D[U] - r'\Sigma^{-1}r.
\end{aligned}
$$

It should be remarked that for computing the best predictor $U^*(X)$ of U we need to know the distribution of the random vector $(X', U)'$. If this vector is normally distributed, then its distribution is uniquely determined by its basic characteristics, the mean value and the covariance matrix, which are the only parameters of any normally distributed random vector.

In the case when we do not know the distribution of the random vector $(X', U)'$, but *the mean value and the covariance matrix of this random vector are known*, we can look for *the best linear predictor $U^*(X)$* of U, which minimizes the MSE in *the class of all linear predictors $\tilde{U}(X)$* of U, which are of the form

$$\tilde{U}(X) = a'X + b; a \in E^n, b \in E^1.$$

The MSE of any predictor $\tilde{U}(X)$ is

$$E[\tilde{U} - U]^2 = D[U] + D[\tilde{U}] + (E[U] - E[\tilde{U}])^2 - 2Cov(U; \tilde{U})$$

and, if $\tilde{U}(X)$ is a linear predictor, then $\tilde{U}(X) = a'X + b$ and

$$E[\tilde{U} - U]^2 = D[U] + a'\Sigma a + (E[U] - a'E[X] - b)^2 - 2a'r,$$

where, as before, $\Sigma = Cov(X)$ and $r = Cov(X; U)$.

Thus to find the best linear predictor $U^*(X)$ of U we have to find

$a^* \in E^n$ and $b^* \in E^1$ such that

$$a^*, b^* = \arg\min_{a,b}\{a'\Sigma a + (E[U] - a'E[X] - b)^2 - 2a'r\}.$$

Setting the partial derivatives equal to zero we get that a^*, b^* are solutions of the equations, in the matrix form,

$$E[U] - a'E[X] - b = 0$$
$$(E[U] - a'E[X] - b)E[X] + \Sigma a - r = 0.$$

The solution is

$$a^* = \Sigma^{-1}r$$

and

$$b^* = E[U] - r'\Sigma^{-1}E[X].$$

Thus the best linear predictor $U^*(X) = a^{*'}X + b^*$ of U is

$$U^*(X) = E[U] + r'\Sigma^{-1}(X - E[X])$$

and we see that this predictor is identical to the best predictor in the case of normality of the random vector $(X', U)'$. We see that the best linear predictor $U^*(X)$ is unbiased for U and has the MSE

$$E[U^* - U]^2 = D[U] - r'\Sigma^{-1}r.$$

The assumption that $E[U]$ and $E[X]$ are known will be replaced, in the following sections of this chapter, by the assumption that they are given by some regression models.

4.2   Predictions in Linear Models

In this section we shall assume that we observe the time series $X(.)$ of the form

$$X(t) = m_\beta(t) + \varepsilon(t) = \sum_{i=1}^{k} \beta_i f_i(t) + \varepsilon(t); t = 1, 2, ...,$$

where the regression parameter $\beta = (\beta_1, ..., \beta_k)'$ can be any vector from E^k and $f_1, ..., f_k$ are given known functions. For time series $\varepsilon(.)$ it is assumed that $E[\varepsilon(t)] = 0$ and $Cov(\varepsilon(s); \varepsilon(t)) = R(s, t); s, t = 1, 2, ...$
Then for any random variable $U = X(n+d)$, where d is a positive integer,

we get

$$E_\beta[U] = f'\beta; \beta \in E^k \text{ and } Cov(U; X(t)) = R(n+d, t); t = 1, 2, ...,$$

where the $k \times 1$ vector $f = (f_1(n+d), f_2(n+d), ..., f_k(n+d))'$ is known.

Next we assume that the covariance function $R(.,.)$ of $X(.)$ is known, but the probability distribution of $X(.)$ will not be specified.

Let $X = (X(1), ..., X(n))'$ be a finite observation of $X(.)$ and let the random vector Z be given by $Z = (X', U)'$. Then the random vector Z is given by the LPRM

$$Z = \begin{pmatrix} X \\ U \end{pmatrix} = \begin{pmatrix} F \\ f' \end{pmatrix} \beta + \varepsilon; \beta \in E^k; E[\varepsilon] = 0; Cov(Z) = \begin{pmatrix} \Sigma & r \\ r' & D[U] \end{pmatrix},$$

where $\Sigma = Cov(X)$ with $\Sigma_{st} = R(s, t)$ and $r = Cov(X, U)$ is an $n \times 1$ vector of covariances between X and U. The components of this vector are $r_t = Cov(X(t); U) = R(t, n+d); t = 1, 2, ..., n$, and are assumed to be known.

In this case when we do not know the distribution of the random vector $(X', U)'$, but the mean value and the covariance matrix of this random vector are known, we can again look for the best linear predictor $U^*(X)$ of U, which minimizes the MSE in the class of all linear predictors $\tilde{U}(X)$ of U, which are of the form

$$\tilde{U}(X) = a'X + b; a \in E^n, b \in E^1.$$

The mean squared error of any linear predictor $\tilde{U}(X)$ is

$$E[\tilde{U} - U]^2 = E_\beta[\tilde{U} - U]^2 = D[U] + a'\Sigma a + (E_\beta[U] - E_\beta[\tilde{U}])^2 - 2a'r$$

and we see that it depends on β, the unknown regression parameters. To avoid these difficulties we should restrict ourselves to the *unbiased linear predictors* for which

$$E_\beta[U] = E_\beta[\tilde{U}] \text{ for all } \beta \in E^k.$$

This condition can be written as

$$E_\beta[\tilde{U}] = a'F\beta + b = f'\beta = E_\beta[U] \text{ for all } \beta \in E^k.$$

Setting $\beta = 0$ we get that $b = 0$ and thus the unbiasedness condition on a is

$$(a'F - f')\beta = 0 \text{ for all } \beta \in E^k$$

and is equivalent to the condition

$$F'a - f = 0.$$

Thus the *best linear unbiased predictor* (BLUP) $U^*(X)$ of U which minimizes the MSE in the class of all linear unbiased predictors of U is given by

$$U^*(X) = \sum_{t=1}^{n} a_t^* X(t),$$

where $a^* = (a_1^*, a_2^*, ..., a_n^*)'$ is a solution of the minimization problem

$$a^* = \arg \min_{\{a:F'a-f=0\}} \{a'\Sigma a - 2a'r\}.$$

This problem can be solved by applying the Lagrange method of indefinite multipliers. Let

$$\phi(\lambda, a) = a'\Sigma a - 2a'r - 2\lambda' F'a + 2\lambda' f.$$

To find

$$a^*(\lambda) = \arg \min_a \phi(\lambda, a)$$

we should solve the equation

$$\frac{\partial \phi(\lambda, a)}{\partial a} \big|_{a^*(\lambda)} = \Sigma a^*(\lambda) - r - F\lambda = 0$$

from which

$$a^*(\lambda) = \Sigma^{-1}(r + F\lambda).$$

Since $U^*(X)$ should be an unbiased predictor, for $a^*(\lambda)$ the condition

$$F'a^*(\lambda) - f = F'\Sigma^{-1}(r + F\lambda) - f = 0$$

must hold. From this condition we get the expression for λ:

$$\lambda = (F'\Sigma^{-1}F)^{-1}(f - F'\Sigma^{-1}r)$$

and thus

$$\begin{aligned} a^* &= \Sigma^{-1}(r + F(F'\Sigma^{-1}F)^{-1}(f - F'\Sigma^{-1}r)) \\ &= \Sigma^{-1}r + \Sigma^{-1}F(F'\Sigma^{-1}F)^{-1}(f - F'\Sigma^{-1}r). \end{aligned}$$

Using this result we can write the following expression for the BLUP $U^*(X)$:

$$\begin{aligned} U^*(X) &= a^{*'}X = f'(F'\Sigma^{-1}F)^{-1}F'\Sigma^{-1}X \\ &\quad + r'\Sigma^{-1}[X - F(F'\Sigma^{-1}F)^{-1}F'\Sigma^{-1}X]. \end{aligned}$$

We get the more convenient form for the BLUP $U^*(X)$ if we use the known result according to which

$$(F'\Sigma^{-1}F)^{-1}F'\Sigma^{-1}X = \beta_\Sigma^*(X)$$

is the BLUE of β and thus

$$U^*(X) = f'\beta_\Sigma^*(X) + r'\Sigma^{-1}(X - F\beta_\Sigma^*(X)),$$

or

$$U^*(X) = E^*[U] + r'\Sigma^{-1}(X - E^*[X]).$$

Thus we can see that the expression for the BLUP $U^*(X)$ of U, in the case when $(X', U)'$ is given by an LPRM, differs from the expression for the best linear predictor in the case when $E[X]$ and $E[U]$ are known, only in such a way that the unknown $E[X]$ and $E[U]$ are replaced by their best estimators $E_\beta^*[U] = f'\beta_\Sigma^*$ and $E_\beta^*[X] = F\beta_\Sigma^*(X)$.

It can also be seen from the expression for the value $U^*(x)$ of the BLUP $U^*(X)$ that it is equal to the prolonged estimated mean value $m^*(n+d)$ equal to $m_{\beta_\Sigma^*}(n+d)$ plus the term $r'\Sigma^{-1}(x - E_\beta^*[X])$ which can be called as corrections from the weighted residuals $x - E_\beta^*[X] = x - F\beta_\Sigma^*(x)$.

In the case when X and U are uncorrelated the BLUP U^* is equal to the prolonged estimated mean value $m_{\beta_\Sigma^*}(n+d)$ and, if the components of X are uncorrelated and have a common variance σ^2, the WELSE β_Σ^* is equal to the OLSE $\hat{\beta}$.

The MSE of the BLUP is

$$E[U^* - U]^2 = D[U] + a^{*'}\Sigma a^* - 2a^{*'}r.$$

Let us use the notation $g = f - F'\Sigma^{-1}r$ and $(F'\Sigma^{-1}F)^{-1} = Cov(\beta_\Sigma^*)$. Then $a^* = \Sigma^{-1}r + \Sigma^{-1}FCov(\beta_\Sigma^*)g$ and thus

$$
\begin{aligned}
E[U^* - U]^2 &= D[U] + (r'\Sigma^{-1} + g'Cov(\beta_\Sigma^*)F'\Sigma^{-1})\Sigma \\
&\quad \times(\Sigma^{-1}r + \Sigma^{-1}FCov(\beta_\Sigma^*)g) \\
&\quad -2r'\Sigma^{-1}r - 2g'Cov(\beta_\Sigma^*)F'\Sigma^{-1}r \\
&= D[U] - r'\Sigma^{-1}r + g'Cov(\beta_\Sigma^*)(F'\Sigma^{-1}F)Cov(\beta_\Sigma^*)g \\
&= D[U] - r'\Sigma^{-1}r + (f - F'\Sigma^{-1}r)'Cov(\beta_\Sigma^*)(f - F'\Sigma^{-1}r),
\end{aligned}
$$

or

$$E[U^* - U]^2 = D[U] - \|r\|_{\Sigma^{-1}}^2 + \|f - F'\Sigma^{-1}r\|_{Cov(\beta_\Sigma^*)}^2.$$

We see from this expression that the MSE of the BLUP is greater than the MSE of the best linear predictor for time series with known mean value.

Let us now consider any linear unbiased estimator $\tilde{\beta}(X)$ of β which can be written as $\tilde{\beta}(X) = LX$, where L is a suitable $k \times n$ matrix, and let

$$\tilde{U}(X) = f'\tilde{\beta}(X) + r'\Sigma^{-1}(X - F\tilde{\beta}(X)).$$

Then it is clear that $\tilde{U}(X)$ is a linear unbiased predictor of U which can be written as

$$\tilde{U}(X) = \tilde{a}'X, \text{ where } \tilde{a} = \Sigma^{-1}r + L(f - F'\Sigma^{-1}r) = \Sigma^{-1}r + L'g.$$

Thus its MSE is

$$
\begin{aligned}
E[\tilde{U} - U]^2 &= D[U] + \tilde{a}'\Sigma\tilde{a} - 2\tilde{a}'r \\
&= D[U] + (r'\Sigma^{-1} + g'L)\Sigma(\Sigma^{-1}r + L'g) \\
&\quad -2r'\Sigma^{-1}r - 2g'Lr \\
&= D[U] - r'\Sigma^{-1}r + g'L\Sigma L'g \\
&= D[U] - r'\Sigma^{-1}r + (f - F'\Sigma^{-1}r)'L\Sigma L'(f - F'\Sigma^{-1}r),
\end{aligned}
$$

or

$$E[\tilde{U} - U]^2 = D[U] - \|r\|^2_{\Sigma^{-1}} + \|f - F'\Sigma^{-1}r\|^2_{Cov(\tilde{\beta})}.$$

It is clear that for all these linear unbiased predictors we get the inequality

$$E[\tilde{U} - U]^2 \geq E[U^* - U]^2,$$

where $U^*(X)$ is the BLUP of U. As a special case of $\tilde{U}(X)$ we can consider the predictor

$$\hat{U}(X) = f'\hat{\beta}(X) + r'\Sigma^{-1}(X - F\hat{\beta}(X)),$$

where $\hat{\beta}(X) = (F'F)^{-1}FX$ is the OLSE of β.

Remark. If $X(.)$ is covariance stationary with a covariance function $R(.)$, then $\Sigma_{n,st} = R(s-t)$ and, if $U = X(n+d)$, $r_t = R(n+d-t)$; $s, t = 1, 2, ..., n$. For $d = 1$ we have $r = (R(n), R(n-1), ..., R(1))'$. It is shown in Brockwell and Davis (1987) that Σ_n are p.d. for every n if $R(0) > 0$ and $\lim_{t\to\infty} R(t) = 0$.

Example 4.2.1. Let us consider the time series

$$X(t) = \sum_{i=1}^{k} \beta_i f_i(t) + \varepsilon(t); t = 1, 2, ...; \beta \in E^k,$$

where $\varepsilon(.)$ is an AR(1) time series with covariance functions

$$R_\nu(t) = \frac{\sigma^2}{1 - \rho^2}\rho^t; t = 0, 1, 2, ...,$$

with $\nu = (\rho, \sigma^2)'$, where $\rho \in (-1, 1)$ and $\sigma^2 \in (0, \infty)$. Then

$$\Sigma_\nu^{-1} = \frac{1}{\sigma^2} \begin{pmatrix} 1 & -\rho & 0 & \ldots & 0 & 0 & 0 \\ -\rho & 1+\rho^2 & -\rho & \ldots & 0 & 0 & 0 \\ \cdot & \cdot & & \ldots & \cdot & \cdot & \cdot \\ 0 & 0 & 0 & \ldots & -\rho & 1+\rho^2 & -\rho \\ 0 & 0 & 0 & \ldots & 0 & -\rho & 1 \end{pmatrix}$$

and, for $d = 1$,

$$r_\nu = \frac{\sigma^2}{1 - \rho^2}(\rho^n, \rho^{n-1}, ..., \rho)'$$

from which we get that

$$\Sigma_\nu^{-1} r_\nu = (0, 0, ..., 0, \rho)'$$

depends only on ρ.

Thus we get that for time series with AR(1) errors the BLUP $X^*(n+1)$ is given by

$$X^*(n+1) = f'\beta_\Sigma^* + \rho[X(n) - (F\beta_\Sigma^*)_n]$$

and it can be seen that in the case when the mean value of $X(.)$ is zero, the BLUP $X^*(n+1)$ depends only on the last observation $X(n)$ of X.

The MSE of the BLUP $X^*(n+1)$ is

$$\begin{aligned} E[X^*(n+1) - X(n+1)]^2 &= \frac{\sigma^2}{1 - \rho^2} - \frac{\sigma^2}{1 - \rho^2}\rho^2 + \|f - \rho f_{(n)}\|^2_{Cov(\beta_\Sigma^*)} \\ &= \sigma^2 + \|f - \rho f_{(n)}\|^2_{Cov(\beta_\Sigma^*)}, \end{aligned}$$

where $f = (f_1(n+1), f_2(n+1), ..., f_k(n+1))'$ and $f_{(n)}$ is the last row of the matrix F.

For a $d-step$ prediction when $U = X(n+d)$ we get

$$r_\nu = \frac{\sigma^2}{1 - \rho^2}(\rho^{n+d-1}, \rho^{n+d-2}, ..., \rho^d)',$$

$$\Sigma_\nu^{-1} r_\nu = (0, 0, ..., 0, \rho^d)'.$$

and thus for a time series with AR(1) errors the BLUP $X^*(n+d), d > 1$, is given by

$$X^*(n+d) = f'\beta_\Sigma^* + \rho^d[X(n) - (F\beta_\Sigma^*)_n]$$

and it can be seen that in the case when the mean value of $X(.)$ is zero, the BLUP $X^*(n+d)$ depends again only on the last observation $X(n)$ of X. The MSE of the BLUP $X^*(n+d)$ is

$$
\begin{aligned}
E[X^*(n+d) - X(n+d)]^2 &= \frac{\sigma^2}{1-\rho^2} - \frac{\sigma^2}{1-\rho^2}\rho^{2d} + \left\| f - \rho^d f_{(n)} \right\|^2_{Cov(\beta_\Sigma^*)} \\
&= \sigma^2\frac{1-\rho^{2d}}{1-\rho^2} + \left\| f - \rho^d f_{(n)} \right\|^2_{Cov(\beta_\Sigma^*)}.
\end{aligned}
$$

If we use the predictor

$$
\hat{X}(n+d) = f'\hat{\beta} + \rho^d[X(n) - (F\hat{\beta})_n],
$$

where $\hat{\beta}$ is the OLSE of β, then

$$
E[\hat{X}(n+d) - X(n+d)]^2 = \sigma^2\frac{1-\rho^{2d}}{1-\rho^2} + \left\| f - \rho^d f_{(n)} \right\|^2_{Cov(\hat{\beta})},
$$

where now $f = (f_1(n+d), f_2(n+d), ..., f_k(n+d))'$.

Example 4.2.2. Let us again consider the time series from the preceding example and let us assume that $X(.)$ has an unknown constant mean value β. Then

$$
\begin{aligned}
F &= (1,1,...,1)', \ (F'F)^{-1} = \frac{1}{n}, \\
\hat{\beta}(X) &= \frac{1}{n}\sum_{t=1}^{n} X(t) \text{ and } Cov(\hat{\beta}) = \frac{1}{n^2}F'\Sigma_\nu F,
\end{aligned}
$$

and

$$
Cov(\hat{\beta}) = \frac{1}{n^2}F'\Sigma_\nu F = \frac{1}{n^2}\left[nR(0) + 2\sum_{t=1}^{n}(n-t)R(t)\right].
$$

Next

$$
\left\| f - \rho^d f_{(n)} \right\|^2_{Cov(\hat{\beta})} = (1-\rho^d)^2\left[\frac{R(0)}{n} + \frac{2}{n}\sum_{t=1}^{n}\left(1-\frac{t}{n}\right)R(t)\right] \to 0 \text{ if } n \to \infty
$$

if $|\rho| < 1$ and thus

$$
\begin{aligned}
\lim_{n\to\infty} E[\hat{X}(n+d) - X(n+d)]^2 &= \sigma^2\frac{1-\rho^{2d}}{1-\rho^2} \\
&= \lim_{n\to\infty} E[X^*(n+d) - X(n+d)]^2.
\end{aligned}
$$

Thus we see that the predictor $\hat{X}(n+d)$ and the BLUP $X^*(n+d)$ are asymptotically equivalent. We say also that they are *adaptive*.

Example 4.2.3. Let us consider again the time series from Example 4.2.1 and let us assume that

$$E_\beta[X(t)] = \beta_1 + \beta_2 t; \beta \in E^2, t = 1, 2, \dots.$$

Then

$$F' = \begin{pmatrix} 1 & 1 & \dots & 1 \\ 1 & 2 & \dots & n \end{pmatrix}, f = \begin{pmatrix} 1 \\ n+d \end{pmatrix},$$

$$f - \rho^d f_{(n)} = \begin{pmatrix} 1 - \rho^d \\ n(1 - \rho^d) + d \end{pmatrix}$$

and the following inequality (which was already used in Section 3.2) holds:

$$\left\| f - \rho^d f_{(n)} \right\|^2_{Cov(\hat{\beta})} \le (f - \rho^d f_{(n)})'(F'F)^{-1}(f - \rho^d f_{(n)}) \left\| \Sigma_\nu \right\|.$$

Next we have

$$\lim_{n \to \infty} (f - \rho^d f_{(n)})' A(f - \rho^d f_{(n)}) = 4(1 - \rho^d)^2,$$

since

$$(F'F)^{-1} = \frac{1}{n} \begin{pmatrix} 2(2n+1) & -6/(n-1) \\ -6/(n-1) & 12/(n^2-1) \end{pmatrix} = \frac{1}{n} A$$

and

$$(f - \rho^d f_{(n)})' A(f - \rho^d f_{(n)}) = \frac{4n+2}{n-1}(1 - \rho^d)^2$$
$$- \frac{12}{n-1}(1 - \rho^d)[n(1 - \rho^d) + d]$$
$$+ \frac{12}{n^2-1}(1 - \rho^d)[n(1 - \rho^d) + d]^2.$$

Using the known result

$$\frac{1}{n^2} \left\| \Sigma \right\|^2 = \frac{R^2(0)}{n} + \frac{2}{n} \sum_{t=1}^{n} \left(1 - \frac{t}{n} \right) R^2(t) \to 0 \text{ for } n \to \infty,$$

we have again

$$0 \le \lim_{n \to \infty} \left\| f - \rho^d f_{(n)} \right\|^2_{Cov(\hat{\beta})}$$
$$\le \lim_{n \to \infty} (f - \rho^d f_{(n)})'(F'F)^{-1}(f - \rho^d f_{(n)}) \left\| \Sigma_\nu \right\| = 0$$

and

$$\lim_{n \to \infty} E[\hat{X}(n+d) - X(n+d)]^2 = \sigma^2 \frac{1 - \rho^{2d}}{1 - \rho^2}$$

$$= \lim_{n \to \infty} E[X^*(n+d) - X(n+d)]^2.$$

Thus we see that also for a linear trend the predictor $\hat{X}(n+d)$ and the BLUP $X^*(n+d)$ are asymptotically equivalent and thus adaptive.

Example 4.2.4. Let us consider the time series

$$X(t) = \sum_{i=1}^{k} \beta_i f_i(t) + \varepsilon(t); t = 1, 2, ...; \beta \in E^k,$$

where $\varepsilon(.)$ is a time series with covariance functions

$$R_\nu(t) = \sigma^2 e^{-\alpha t}; t = 0, 1, 2, ...,$$

with $\nu = (\sigma^2, \alpha)'$, where $\sigma^2 \in (0, \infty)$ and $\alpha \in (0, \infty)$. Then

$$\Sigma_\nu^{-1} = \frac{1}{\sigma^2(1 - e^{-2\alpha})} \begin{pmatrix} 1 & -e^{-\alpha} & 0 & \cdots & 0 & 0 \\ -e^{-\alpha} & 1 + e^{-2\alpha} & -e^{-\alpha} & \cdots & 0 & 0 \\ \cdot & \cdot & \cdot & \cdots & \cdot & \cdot \\ 0 & 0 & 0 & \cdots & 1 + e^{-\alpha 2} & -e^{-\alpha} \\ 0 & 0 & 0 & \cdots & -e^{-\alpha} & 1 \end{pmatrix}$$

and, for $d = 1$,

$$r_\nu = \sigma^2(e^{-\alpha n}, e^{-\alpha(n-1)}, ..., e^{-\alpha})',$$

$$\Sigma_\nu^{-1} r_\nu = (0, 0, ..., 0, e^{-\alpha})'.$$

Thus we get that, for these time series, the BLUP $X^*(n+1)$ is given by

$$X^*(n+1) = f'\beta_\Sigma^* + e^{-\alpha}[X(n) - (F\beta_\Sigma^*)_n]$$

and it can again be seen that in the case when the mean value of $X(.)$ is zero, the BLUP $X^*(n+1)$ depends only on the last observation $X(n)$ of X.

For a d−step prediction when $U = X(n+d)$ we get

$$\Sigma_\nu^{-1} r_\nu = (0, 0, ..., 0, e^{-\alpha d})'.$$

Thus we get that the BLUPs $X^*(n+d), d > 1$, are given by

$$X^*(n+d) = f'\beta_\Sigma^* + e^{-\alpha d}[X(n) - (F\beta_\Sigma^*)_n]$$

and it can be seen that in the case when the mean value of $X(.)$ is zero, the BLUPs $X^*(n+d)$ depend again only on the last observation $X(n)$ of X.

The MSE of the BLUP $X^*(n+d)$ is

$$
\begin{aligned}
E[X^*(n+d) - X(n+d)]^2 &= \sigma^2 - \sigma^2 e^{-\alpha 2d} + \left\| f - e^{-\alpha d} f_{(n)} \right\|^2_{Cov(\beta^*_\Sigma)} \\
&= \sigma^2(1 - e^{-\alpha 2d}) + \left\| f - e^{-\alpha d} f_{(n)} \right\|^2_{Cov(\beta^*_\Sigma)}.
\end{aligned}
$$

It can easily be seen that for the mean values considered in Examples 4.2.2 and 4.2.3 the predictors $\hat{X}(n+d)$ and the BLUPs $X^*(n+d)$ are adaptive, because they are asymptotically equivalent.

Example 4.2.5. Let us consider the time series

$$
X(t) = \sum_{i=1}^{k} \beta_i f_i(t) + \varepsilon(t); t = 1, 2, ...; \beta \in E^k,
$$

with covariance functions

$$
R_\nu(t) = \sigma^2 e^{-\alpha t} \cos \beta t; t = 0, 1, 2, ...,
$$

where $\nu = (\sigma^2, \alpha, \beta)' \in \Upsilon = (0, \infty) \times (0, \infty) \times [0, \pi]$.

Since the exact expression for Σ_ν^{-1} is not known, we shall use numerical results for the given values of parameter ν derived in Example 3.4.6. Let ν be equal to $\nu_0 = (1, 2, \pi/4)'$ and let $n = 10$. Then

$$
\Sigma_{\nu_0}^{-1} = \begin{pmatrix}
1 & -.01 & 0 & \cdots & 0 & 0 & 0 \\
-.01 & 1 & -.01 & \cdots & 0 & 0 & 0 \\
0 & -.01 & 1 & \cdots & 0 & 0 & 0 \\
. & . & . & \cdots & . & . & . \\
0 & 0 & 0 & \cdots & 1 & -.01 & 0 \\
0 & 0 & 0 & \cdots & -.01 & 1 & -.01 \\
0 & 0 & 0 & \cdots & 0 & -.01 & 1
\end{pmatrix}
$$

and, for $d = 1$, after some algebra,

$$
\Sigma_{\nu_0}^{-1} r_{\nu_0} = (0, 0, 0, 0, 0, 0, 0, 0, 0, .0579)'.
$$

Thus, for this value of parameter ν we get that the BLUP $X^*(n+1)$ depends only on the last observation $X(n)$ of X.

For the case when $\nu = \nu_1 = (1, 0.1, \pi/4)'$ and $n = 10$ we get

$$
\Sigma_{\nu_1}^{-1} = \begin{pmatrix}
3.5 & -3.1 & 1.7 & \cdots & .1 & 0 & 0 \\
-3.1 & 6.3 & -4.6 & \cdots & -.1 & .1 & 0 \\
1.7 & -4.6 & 7.1 & \cdots & .1 & -.1 & .1 \\
. & . & . & \cdots & . & . & . \\
.1 & -.1 & .1 & \cdots & 7.1 & -4.6 & 1.7 \\
0 & .1 & -.1 & \cdots & -4.6 & 6.3 & -3.1 \\
0 & 0 & .1 & \cdots & 1.7 & -3.1 & 3.5
\end{pmatrix}
$$

and, for $d = 1$,

$$r_\nu = \sigma^2 (e^{-\alpha n} \cos \beta n, e^{-\alpha(n-1)} \cos \beta(n-1), ..., e^{-\alpha} \cos \beta)',$$

from which, for $\nu = \nu_1$, we get that

$$
\begin{aligned}
r_{\nu_1} &= (0, .29, .45, .35, 0, -.43, -.67, -.52, 0, .64)', \\
r'_{\nu_1} \Sigma_{\nu_1}^{-1} &= (-.01, -.01, 0, -.02, -.01, -.01, -.12, -.16, -.46, .86)
\end{aligned}
$$

and thus the BLUP is a linear combination of all the components of X. It is easy to find that

$$r'_{\nu_1} \Sigma_{\nu_1}^{-1} r_{\nu_1} = .712.$$

Using this result we can write that

$$
\begin{aligned}
E_{\nu_1}[X^*(n+1) - X(n+1)]^2 &= R_{\nu_1}(0) - r'_{\nu_1} \Sigma_{\nu_1}^{-1} r_{\nu_1} \\
&\quad + \left\| f - F' \Sigma_{\nu_1}^{-1} r_{\nu_1} \right\|^2_{Cov(\beta^*_{\nu_1})} \\
&= .288 + \left\| f - F' \Sigma_{\nu_1}^{-1} r_{\nu_1} \right\|^2_{Cov(\beta^*_{\nu_1})}.
\end{aligned}
$$

Example 4.2.6. Let $X(.)$ be a homogeneous Poisson process with parameter λ. Then the random variables $X(t)$ have Poisson distribution with parameter λt and thus

$$E_\lambda[X(t)] = \lambda t; \lambda \in (0, \infty).$$

The covariance function of a homogeneous Poisson process $X(.)$ is given by

$$R_\lambda(s, t) = \lambda \min(s; t); \lambda \in (0, \infty).$$

If X is a finite observation of $X(.)$, then we can write

$$E_\lambda[X] = F\lambda,$$

where $F = (1, 2, ..., n)'$ and

$$\Sigma_\lambda = Cov_\lambda(X) = \lambda \begin{pmatrix} 1 & 1 & ... & 1 \\ 1 & 2 & ... & 2 \\ . & . & ... & . \\ 1 & 2 & ... & n \end{pmatrix}.$$

The inverse Σ_λ^{-1} of the covariance matrix Σ_λ is given by

$$\Sigma_\lambda^{-1} = \frac{1}{\lambda} \begin{pmatrix} 2 & -1 & 0 & \cdots & 0 & 0 & 0 \\ -1 & 2 & -1 & \cdots & 0 & 0 & 0 \\ \cdot & \cdot & \cdot & \cdots & \cdot & \cdot & \cdot \\ 0 & 0 & 0 & \cdots & -1 & 2 & -1 \\ 0 & 0 & 0 & \cdots & 0 & -1 & 1 \end{pmatrix},$$

from which $F'\Sigma_\lambda^{-1} = 1/\lambda\,(0,0,...,1)'$ and $F'\Sigma_\lambda^{-1}F = n/\lambda$.

From these expressions we get the expression for the BLUE λ^* of λ:

$$\lambda^*(X) = (F'\Sigma_\lambda^{-1}F)^{-1}F'\Sigma_\lambda^{-1}X = \frac{X(n)}{n}.$$

Its variance is

$$D_\lambda[\lambda^*(X)] = \frac{\lambda}{n}.$$

Next we get, for $U = X(n+d)$, that

$$r_\lambda = Cov_\lambda(X;U) = \lambda(1,2,...,n)' = \lambda F$$

for any $d \geq 1$ and thus, for all λ,

$$r_\lambda'\Sigma_\lambda^{-1} = (0,0,...,1)'.$$

We get the BLUP $X^*(n+d), d \geq 1$, in the form

$$X^*(n+d) = \lambda^*(n+d) + (X(n) - \lambda^*n) = \lambda^*(n+d),$$

since $X(n) - \lambda^*n = 0$ and we see that the BLUP $X^*(n+d)$ is simply a prolonged estimated mean value, despite the fact that $r_\lambda'\Sigma_\lambda^{-1} \neq 0$. We remark that the residuals $X(t) - \lambda^*t = X(t) - (X(n)/n)\,t; t = 1,2,...,n-1$, are not necessarily equal to zero.

The MSE of the BLUP $X^*(n+d)$ is

$$\begin{aligned} E[X^*(n+d) - X(n+d)]^2 &= \lambda(n+d) - \lambda n + \|n+d-n\|^2_{Cov(\lambda^*)} \\ &= \lambda d + d^2\frac{\lambda}{n}. \end{aligned}$$

and

$$\lim_{n\to\infty} E[X^*(n+d) - X(n+d)]^2 = \lambda d.$$

Instead of the BLUE λ^* we can use the OLSE $\hat{\lambda}$. It is given by

$$\hat{\lambda}(X) = \frac{6}{n(n+1)(2n+1)} \sum_{t=1}^{n} tX(t)$$

and its variance is

$$D_\lambda[\hat{\lambda}(X)] = \left(\frac{6}{n(n+1)(2n+1)}\right)^2 F'\Sigma_\lambda F.$$

Next we have

$$
\begin{aligned}
D_\lambda[\hat{\lambda}(X)] &= \lambda \left(\frac{6}{n(n+1)(2n+1)}\right)^2 \sum_{s=1}^{n}\sum_{t=1}^{n} st \min(s;t) \\
&\leq \lambda \left(\frac{6}{n(n+1)(2n+1)}\right)^2 n \left(\sum_{t=1}^{n} t\right)^2 \\
&= \lambda \left(\frac{6}{n(n+1)(2n+1)}\right)^2 \frac{1}{4} n^3 (n+1)^2
\end{aligned}
$$

and we can deduce that the OLSE $\hat{\lambda}(X)$ is a consistent unbiased estimator. The analogue of the BLUP, the predictor $\hat{X}(n+d)$, is given by

$$\hat{X}(n+d) = \hat{\lambda}(n+d) + (X(n) - \hat{\lambda}n)$$

and has the MSE

$$E[\hat{X}(n+d) - X(n+d)]^2 = \lambda d + d^2 D_\lambda[\hat{\lambda}].$$

It is clear that $D_\lambda[\hat{\lambda}] \geq D_\lambda[\lambda^*]$. It can be computed that for $n = 5$ we get $D_\lambda[\hat{\lambda}] = 0.221\lambda$ and $D_\lambda[\lambda^*] = 0.200\lambda$. For $n = 20$ we get $D_\lambda[\hat{\lambda}] = 0.058\lambda$ and $D_\lambda[\lambda^*] = 0.050\lambda$ and for $n = 90$, $D_\lambda[\hat{\lambda}] = 0.013\lambda$ and $D_\lambda[\lambda^*] = 0.011\lambda$.

We can see that these variances are practically the same and thus the MSEs of the BLUP $X^*(n+d)$ and of $\hat{X}(n+d)$ are also practically the same.

Example 4.2.7. Let us consider the time series

$$X(t) = \sum_{i=1}^{k} \beta_i f_i(t) + \varepsilon(t); t = 1, 2, ...; \beta \in E^k,$$

where $\varepsilon(.)$ are the AR(p) time series with such parameters $a = (a_1, ..., a_p)'$ for which $\varepsilon(.)$ and thus also $X(.)$ are covariance stationary. To compute the BLUP $X^*(n+d)$ we need to know the vector $\Sigma^{-1}r$ which can be computed directly for an AR(1) time series, when the covariance function of $X(.)$ is given explicitly. This approach cannot be used for an AR(p) time series with $p > 1$, since in general we have no explicit expression for $R(.)$. But for an $AR(p)$ time series the BLUP $X^*(n+d)$ can be derived directly by using the following approach.

Since

$$\Sigma_{st} = Cov(X(s); X(t)) = E[\varepsilon(s)\varepsilon(t)]$$

and also

$$r_t = Cov(X(t); X(n+d)) = E[\varepsilon(t)\varepsilon(n+d)]; s, t = 1, 2, ..., n,$$

the vector $\Sigma^{-1}r$ can be identified with the vector which determines the BLUP $\varepsilon^*(n+d)$ for the $AR(p)$ time series $\varepsilon(.)$ with mean value zero defined by

$$\varepsilon(t) = \sum_{i=1}^{p} a_i \varepsilon(t-i) + \varpi(t); t = ..., -2, -1, 0, 1, 2, ...,$$

where $\varpi(.)$ is a white noise with variance σ^2 for which

$$E[\varpi(s)\varpi(t)] = 0 \text{ for all } s \neq t.$$

Next we have

$$E[\varepsilon(s)\varpi(t)] = 0 \text{ for all } s < t.$$

Let $n > p$, let $\varepsilon = (\varepsilon(1), ..., \varepsilon(n))'$, and let $\varepsilon(n+1)$ be predicted. Then we can write

$$\varepsilon(n+1) = \sum_{i=1}^{p} a_i \varepsilon(n+1-i) + \varpi(n+1)$$

and thus

$$r_t = E[\varepsilon(n+1)\varepsilon(t)] = E\left[\sum_{i=1}^{p} a_i \varepsilon(n+1-i)\varepsilon(t)\right]; t = 1, 2, ..., n,$$

from which it can be derived that the BLUP for $\varepsilon(n+1)$ is identical to the BLUP for $\sum_{i=1}^{p} a_i \varepsilon(n+1-i)$:

$$\varepsilon^*(n+1) = \left(\sum_{i=1}^{p} a_i \varepsilon(n+1-i)\right)^* = \sum_{i=1}^{p} a_i \varepsilon(n+1-i),$$

since the last sum is a linear combination of $\varepsilon(1), ..., \varepsilon(n)$ and thus is identical to the BLUP.

Next we have

$$E[\varepsilon^*(n+1) - \varepsilon(n+1)]^2 = E[\varpi^2(n+1)] = \sigma^2.$$

Thus we get that

$$\Sigma^{-1}r = (0, ..., 0, a_p, ..., a_1)'$$

and the BLUP $X^*(n+1)$ is given by

$$X^*(n+1) = f'\beta_\Sigma^* + \sum_{i=1}^{p} a_i[X(n+1-i) - (F\beta_\Sigma^*)_{n+1-i}]$$

and

$$E[X^*(n+1) - X(n+1)]^2 = \sigma^2 + \left\| f - F'\Sigma^{-1}r \right\|_{Cov(\beta_\Sigma^*)}^2$$

To solve the problem of d–step prediction the following approach can be used. We can write, for $\varepsilon(n+d)$, the expression

$$\varepsilon(n+d) = \sum_{i=1}^{p} a_i\varepsilon(n+d-i) + \varpi(n+d)$$

and, if we express every $\varepsilon(n+d-i); i = 1, 2, ..., p$, using the equation of autoregression as a sum of the previous values $\varepsilon(.)$, we can write

$$\varepsilon(n+d) = \sum_{i=1}^{p} b_{id}\varepsilon(n+1-i) + \sum_{i=1}^{d} c_i\varpi(n+i),$$

where $b_{id}; i = 1, 2, ..., p$ and $c_i; i = 1, 2, ..., d$, are some constants. By analogy, as for the case when $p = 1$, we have

$$E\left[\sum_{i=1}^{d} c_i\varpi(n+i)\varepsilon(t) \right] = 0 \text{ for } t = 1, 2, ..., n.$$

Thus

$$\varepsilon^*(n+d) = \sum_{i=1}^{p} b_{id}\varepsilon(n+1-i) \text{ if } p < n$$

and

$$E[\varepsilon^*(n+d) - \varepsilon(n+d)]^2 = E\left[\sum_{i=1}^{d} c_i\varpi(n+i) \right]^2 = \sigma^2 \sum_{i=1}^{d} c_i^2.$$

If, for example, $d = 2$ and $\varepsilon(n + 2)$ can be predicted, we write

$$
\begin{aligned}
\varepsilon(n + 2) &= a_1\varepsilon(n + 1) + a_2\varepsilon(n) + \varpi(n + 2) \\
&= a_1[a_1\varepsilon(n) + a_2\varepsilon(n - 1) + \varpi(n + 1)] + a_2\varepsilon(n) + \varpi(n + 2) \\
&= (a_1^2 + a_2)\varepsilon(n) + a_1 a_2\varepsilon(n - 1) + a_1\varpi(n + 1)] + \varpi(n + 2),
\end{aligned}
$$

from which

$$
\varepsilon^*(n + 2) = (a_1^2 + a_2)\varepsilon(n) + a_1 a_2\varepsilon(n - 1)
$$

and the MSE of this estimator is given by

$$
E[\varepsilon^*(n + 2) - \varepsilon(n + 2)]^2 = \sigma^2(a_1^2 + 1).
$$

In this case we have

$$
\Sigma^{-1} r = (0, ..., 0, a_1^2 + a_2, a_1 a_2)'.
$$

Thus the BLUP $X^*(n + d), d \geq 1$, for a time series with AR(p) errors is given by

$$
X^*(n + d) = f'\beta_\Sigma^* + \sum_{i=1}^{p} b_{id}[X(n + 1 - i) - (F\beta_\Sigma^*)_{n+1-i}]
$$

and the correction from the residuals term is a linear combination of only the last p values of the observation X.

Since in general the covariance matrix Σ of X is not known, and thus we don't know the BLUE β_Σ^* of β, it is possible, instead of $X^*(n + d)$, to use the linear predictor $\hat{X}(n + d)$ given by

$$
\hat{X}(n + d) = f'\hat{\beta} + \sum_{i=1}^{p} b_{id}[X(n + 1 - i) - (F\hat{\beta})_{n+1-i}],
$$

where $\hat{\beta}$ is the OLSE of β.

The MSE of this predictor is

$$
MSE[\hat{X}(n + d)] = \sigma^2 \sum_{i=1}^{d} c_i^2 + \left\| f - F'\Sigma^{-1} r \right\|_{Cov(\hat{\beta})}^2.
$$

But, since $Cov(\hat{\beta}) = Cov_\Sigma(\hat{\beta})$ depends on Σ, it is not possible to give an exact expression for the $MSE[\hat{X}(n + d)]$ as in the case when $p = 1$.

The problem the of prediction of time series which are given by MA(q), ARMA(p,q) and ARIMA(p,d,q) models is considered in many books dealing with the Box-Jenkins methodology. The books by Box-Jenkins (1976),

Brockwell and Davis (1987), (1996), Harvey (1994), (1996), and many others can be recommended as a good guide through the problems of the
prediction of time series by using the methodology of Box and Jenkins. It
should also be remarked that many packages of statistical software con-
tain programs for thepredictions of time series by using ARMA(p,q) and
ARIMA(p,d,q) models. The problem of choosing the right model for real
data is usually left to the user and thus different users, different statisti-
cians, can get different results for predictions, using the same methodology.
But this is also true in the case when we are using regression models for
modeling time series data. The conclusion is that it is recommended to use
more models and methods for the data, to compute more predictions, and
thus get a more objective picture of possible future values of an observed
times series.

4.3 Model Choice and Predictions

For any time series $X(.)$ we can choose different LRMs which influence the
predicted values of this time series. Now we shall study the influence of a
choice of models for an observed time series on their predictions.

Let us assume that for a time series $X(.)$ we use *a finite discrete sprectrum
model*

$$X(t) = \sum_{i=1}^{l} Y_i f_i(t); t = 1, 2, ...,$$

where $f_i(.); i = 1, 2, ..., l$, are known functions and where $Y = (Y_1, ..., Y_l)'$ is
a random vector with

$$E[Y] = 0, Cov(Y) = diag(\sigma_i^2).$$

Under these assumptions the covariance functions $R_\nu(.,.)$ of $X(.)$ are
given by

$$R_\nu(s, t) = \sum_{i=1}^{l} \sigma_i^2 f_i(s) f_i(t); s, t = 1, 2, ...$$

Then for a finite observation $X = (X(1), ..., X(n))'$ of time series $X(.)$
we have the model

$$X = FY,$$

where $F = (f_1, ..., f_l), f_i = (f_i(1), ..., f_i(n))'; i = 1, 2, ..., l$. The covariance
matix $Cov(X)$ of X is given by

$$Cov(X) = \Sigma_\nu = \sum_{i=1}^{l} \sigma_i^2 f_i f_i' = \sum_{i=1}^{l} \sigma_i^2 V_i, r(V_i) = 1; i = 1, 2, ..., l.$$

If $l < n$, then $r(\Sigma_\nu) \le n$, and thus $\Sigma_\nu; \nu = (\sigma_1^2, ..., \sigma_l^2)' \in (0, \infty)^l = \Upsilon$, are singular matrices which in general are not stationary.

Any realization $x(.)$ of $X(.)$ is given by

$$x(t) = \sum_{i=1}^{l} y_i f_i(t); t = 1, 2, ..., n,$$

and in the case when $y_i; i = 1, 2, ..., l$, are identifiable from the data $x(t) = \sum_{i=1}^{l} y_i f_i(t); t = 1, 2, ..., n$, we can use for $U = X(n+1)$ the predictor

$$X^*(n+1) = \sum_{i=1}^{l} Y_i f_i(n+1).$$

For this predictor we have

$$MSE_\nu[X^*(n+1)] = E[X^*(n+1) - X(n+1)]^2 = 0$$

and thus $X^*(n+1)$ is the best predictor for $U = X(n+1)$.

Remark. The best prediktor $X^*(n+1)$ cannot be computed from the expression

$$X^*(n+1) = r_\nu' \Sigma_\nu^{-1} X$$

for the BLUP, because the covariance matrices Σ_ν are singular.

Random variables $Y_i; i = 1, 2, ..., l$, are identifiable from X if the design matrix F is a full rank matrix, that is when the vectors $f_1, ..., f_l$ are linearly independent. In this case we have

$$Y = (F'F)^{-1} F' X$$

and

$$X^*(n+1) = \sum_{i=1}^{l} Y_i f_i(n+1) = f'(F'F)^{-1} F' X,$$

where $f = (f_1(n+1), ..., f_l(n+1))'$.

Any finite discrete sprectrum model with identifiable random variables $Y_i; i = 1, 2, ..., l$, can be considered as an example of a *purely deterministic* model in which

$$MSE[X^*(n+d)] = E[X^*(n+d) - X(n+d)]^2 = 0$$

for every $d > 0$.

The *model of random line* given by

$$X(t) = Y_1 + Y_2 t; ; t = 1, 2, ...,$$

in which $Y_i; i = 1, 2$, are identifiable if $n \geq 2$ is an example of a purely deterministic model.

In some cases we can also get singular covariance matrices which are stationary, as is shown in the next example.

Example 4.3.1. Let

$$X(t) = \sum_{i=1}^{k} (Y_i \cos \lambda_i t + Z_i \sin \lambda_i t); t = 1, 2, ...,$$

where $\lambda_1, \lambda_2 \in [-\pi, \pi]$ and where

$$
\begin{aligned}
D[Y_i] &= D[Z_i] = \sigma_i^2, \\
Cov(Y_i; Z_j) &= 0 \text{ for } i \neq j.
\end{aligned}
$$

Then the covariance function $R_\nu(.,.)$ of $X(.)$ is given by

$$
\begin{aligned}
R_\nu(s, t) &= \sum_{i=1}^{k} \sigma_i^2 (\cos \lambda_i s \cos \lambda_i t + \sin \lambda_i s \sin \lambda_i t) \\
&= \sum_{i=1}^{k} \sigma_i^2 \cos \lambda_i (s - t)
\end{aligned}
$$

and we can see that $X(.)$ is covariance stationary. For the covariance matrix $Cov(X)$ of $X = (X(1), ..., X(n))'$, we can write

$$
\begin{aligned}
Cov(X) &= \Sigma_\nu = \sum_{i=1}^{k} \sigma_i^2 f_i f_i' + \sum_{i=1}^{k} \sigma_i^2 g_i g_i' \\
&= \sum_{i=1}^{k} \sigma_i^2 (W_i + U_i) \\
&= \sum_{i=1}^{k} \sigma_i^2 V_i, r(V_i) = 2; i = 1, 2, ..., k,
\end{aligned}
$$

and for $n > l = 2k$, the matrices $\Sigma_\nu; \nu \in \Upsilon$, are singular. But for $n = l$ the matices Σ_ν can be nonsingular.

If we assume that $\sigma_i^2 = D[Y_i] \neq \kappa_i^2 = D[Z_i]$ for $i = 1, 2, ...k$, then $X(.)$ is not covariance stationary and its covariance function will be

$$R_\nu(s, t) = \sum_{i=1}^{k} (\sigma_i^2 \cos \lambda_i s \cos \lambda_i t + \kappa_i^2 \sin \lambda_i s \sin \lambda_i t); s, t = 1, 2, ...$$

In this case we can again write

$$
\begin{aligned}
Cov(X) &= \Sigma_\nu = \sum_{i=1}^{k}(\sigma_i^2 f_i f_i' + \kappa_i^2 g_i g_i') \\
&= \sum_{i=1}^{k}(\sigma_i^2 W_i + \kappa_i^2 U_i) \\
&= \sum_{i=1}^{l} \nu_i V_i, r(V_i) = 1; i = 1, 2, ..., l,
\end{aligned}
$$

and for $n > l$ the matrices $\Sigma_\nu; \nu \in \Upsilon$ will not be regular.

The random variables $Y_i; i = 1, 2, ..., l$, are identifiable for n even, $n > k$, if $f_i(t) = \cos \lambda_i t$, or $f_i(t) = \sin \lambda_i t; i = 1, 2, ..., k$, $\lambda_1, \lambda_2, ..., \lambda_k \in [-\pi, \pi]$, are some of the Fourier frequencies defined as $\lambda_j = (2\pi/n) j; j = 1, 2, ..., n/2$. In this case we have

$$(f_i, f_j) = \frac{n}{2} \text{ if } i = j, \text{ and } (f_i, f_j) = 0 \text{ for } i \neq j,$$

the vectors $f_i; i = 1, 2, ..., 4$, are linearly independent and the matrix $F'F$ is diagonal

$$
F'F = \begin{pmatrix}
n/2 & 0 & \cdots & 0 \\
0 & n/2 & \cdots & 0 \\
\cdot & \cdot & \cdots & \cdot \\
0 & 0 & \cdots & n/2
\end{pmatrix}.
$$

Example 4.3.2. Let us consider a time series $X(.)$ given by

$$
\begin{aligned}
X(t) &= Y_1 \cos \lambda_1 t + Y_2 \sin \lambda_1 t + Y_3 \cos \lambda_2 t + Y_4 \sin \lambda_2 t; t = 1, 2, ..., \\
D[Y_1] &= D[Y_2] = \sigma_1^2, D[Y_3] = D[Y_4] = \sigma_2^2, Cov(Y_i; Y_j) = 0 \text{ for } i \neq j.
\end{aligned}
$$

We know that this time series is covariance stationary with

$$R_\nu(t) = \sigma_1^2 \cos \lambda_1 t + \sigma_2^2 \cos \lambda_2 t; t = 0, 1, 2, ...$$

If $n = 4, \lambda_1 = 0.1\pi, \lambda_2 = 0.8\pi, \sigma_1^2 = 2$ and $\sigma_2^2 = 0.1$, then, since $n = l = 4$, the time series $X(.)$ is purely deterministic, the covariance matrix Σ of X is nonsingular, and we can compute that

$$
\Sigma = \begin{pmatrix}
2.10 & 1.82 & 1.64 & 1.20 \\
1.82 & 2.10 & 1.82 & 1.64 \\
1.64 & 1.82 & 2.10 & 1.82 \\
1.20 & 1.64 & 1.82 & 2.10
\end{pmatrix}, \Sigma^{-1} r = \begin{pmatrix}
0.54 \\
1.21 \\
1.65 \\
1.82
\end{pmatrix}, r'\Sigma^{-1}r = 2.10.
$$

Thus

$$X^*(5) = r'\Sigma^{-1}X$$
$$= 0.54X(1) + 1.21X(2) + 1.65X(3) + 1.82X(4)$$

and

$$MSE_\nu[X^*(5)] = D[X(5)] - r'_\nu\Sigma_\nu^{-1}r_\nu$$
$$= 2.10 - 2.10 = 0$$

and we obtain the expected result, since $X(.)$ is a purely deterministic time series.

In real situations there are only a few cases when we can use, for real data, a finite discrete spectrum model. But in many practical applications of time series theory we can use for an observed time series $X(.)$ *a finite discrete sprectrum with an additive white noise model* given by

$$X(t) = \sum_{i=1}^{l} Y_i f_i(t) + w(t); t = 1, 2, ...,$$
$$E[Y] = 0, Cov(Y) = diag(\sigma_i^2),$$

where $w(.)$ is a white noise uncorrelated with $Y = (Y_1, Y_2, ..., Y_l)'$ and with variance $D[w(t)] = \sigma^2$:

$$Cov(Y, w) = 0, D[w(t)] = \sigma^2.$$

Under these assumptions the covariance functions $R_\nu(., .)$ of $X(.)$ are given by

$$R_\nu(s, t) = \sigma^2\delta_{s,t} + \sum_{i=1}^{l} \sigma_i^2 f_i(s)f_i(t); s, t = 1, 2, ...$$

For this model we get, for a finite observation $X = (X(1), ..., X(n))'$ of $X(.)$, the model

$$X = FY + w,$$

where the $n \times l$ matrix $F = (f_1...f_l)$ has columns $f_i = (f_i(1), ..., f_i(n))'; i = 1, 2, ..., l$. In this model $E[X] = 0$ and the covariance matices $\Sigma_\nu; \nu \in \Upsilon$, of X are p.d. and are given by

$$\Sigma_\nu = \sum_{i=0}^{l} \sigma_i^2 V_i = \sigma^2 I_n + \sum_{i=1}^{l} \sigma_i^2 V_i,$$

where $V_i = f_i f_i'$, $r(V_i) = 1; i = 1, 2, ..., l$.

According to the classical theory the best linear predictor $X^*(n+1)$ of $X(n+1)$ is given by

$$X^*(n+1) = r_\nu' \Sigma_\nu^{-1} X,$$

where $r_\nu = Cov_\nu(X; X(n+1))$ and

$$
\begin{aligned}
MSE_\nu[X^*(n+1)] &= E_\nu[X^*(n+1) - X(n+1)]^2 \\
&= D_\nu[X(n+1)] - r_\nu' \Sigma_\nu^{-1} r_\nu.
\end{aligned}
$$

To compute the matrix Σ_ν^{-1} the following lemma is fundamental.

Lemma 4.3.1. *For any p.d. symmetric $n \times n$ matrix A and for any $n \times 1$ vectors u, v we have*

$$(A + uv')^{-1} = A^{-1} - \frac{A^{-1} uv' A^{-1}}{1 + v' A^{-1} u}.$$

Proof. By direct computation.

From this lemma we easily get the expression

$$
\begin{aligned}
(\sigma^2 I_n + \sigma_1^2 f_1 f_1')^{-1} &= \frac{1}{\sigma^2} \left(I_n - \frac{\sigma_1^2/\sigma^2}{1 + (\sigma_1^2/\sigma^2) \|f_1\|^2} f_1 f_1' \right) \\
&= \frac{1}{\sigma^2} \left(I_n - \frac{1}{\sigma^2/\sigma_1^2 + \|f_1\|^2} f_1 f_1' \right)
\end{aligned}
$$

and the expression

$$\Sigma_\nu^{-1} = \left(\sigma^2 I_n + \sum_{i=1}^{l} \sigma_i^2 f_i f_i' \right)^{-1} = \frac{1}{\sigma^2} \left(I_n - \sum_{i=1}^{l} \frac{1}{\sigma^2/\sigma_i^2 + \|f_i\|^2} f_i f_i' \right)$$

which holds under the assumpton that $f_i; i = 1, 2, ..., l$, are orthogonal vectors, which means $(f_i, f_j) = 0$ for $i \neq j$. We can see that this expression is also true by direct computation.

Using this result we get, for orthogonal $f_i; i = 1, 2, ..., l$,

$$
\begin{aligned}
r_\nu' \Sigma_\nu^{-1} r_\nu &= \frac{1}{\sigma^2} \left(\sum_{j=1}^{l} \sigma_j^2 f_j(n+d) f_j \right)' \left(I_n - \sum_{i=1}^{l} \frac{1}{\sigma^2/\sigma_i^2 + \|f_i\|^2} f_i f_i' \right) \\
&\quad \times \left(\sum_{k=1}^{l} \sigma_k^2 f_k(n+d) f_k \right)
\end{aligned}
$$

$$= \frac{1}{\sigma^2} \left(\sum_{i=1}^{l} \sigma_i^4 f_i^2(n+d) \, \|f_i\|^2 - \sum_{i=1}^{l} \sigma_i^4 f_i^2(n+d) \frac{\|f_i\|^4}{\sigma^2/\sigma_i^2 + \|f_i\|^2} \right)$$

$$= \frac{1}{\sigma^2} \sum_{i=1}^{l} \sigma_i^4 f_i^2(n+d) \, \|f_i\|^2 \left(1 - \frac{\|f_i\|^2}{\sigma^2/\sigma_i^2 + \|f_i\|^2} \right)$$

$$= \sum_{i=1}^{l} \sigma_i^2 f_i^2(n+d) \frac{\|f_i\|^2}{\sigma^2/\sigma_i^2 + \|f_i\|^2}.$$

Next we have

$$D_\nu[X(n+d)] = \sigma^2 + \sum_{i=1}^{l} \sigma_i^2 f_i^2(n+d)$$

and thus

$$\begin{aligned}
MSE_\nu[X^*(n+d)] &= D_\nu[X(n+d)] - r_\nu' \Sigma_\nu^{-1} r_\nu \\
&= \sigma^2 + \sum_{i=1}^{l} \sigma_i^2 f_i^2(n+d) \\
&\quad - \sum_{i=1}^{l} \sigma_i^2 f_i^2(n+d) \frac{\|f_i\|^2}{\sigma^2/\sigma_i^2 + \|f_i\|^2} \\
&= \sigma^2 + \sum_{i=1}^{l} \sigma_i^2 f_i^2(n+d) \left(1 - \frac{\|f_i\|^2}{\sigma^2/\sigma_i^2 + \|f_i\|^2} \right) \\
&= \sigma^2 \left(1 + \sum_{i=1}^{l} \frac{1}{\sigma^2/\sigma_i^2 + \|f_i\|^2} f_i^2(n+d) \right).
\end{aligned}$$

Now we shall compare a finite discrete sprectrum with an additive white noise model, with the second model, *a linear regression, with an additive white noise model,* given by

$$X(t) = \sum_{i=1}^{l} \beta_i f_i(t) + w(t); t = 1, 2, ...; \beta = (\beta_1, ..., \beta_l)' \in E^l,$$

where $w(.)$ is a white noise with variance $D[w(t)] = \sigma^2$.

It is clear that if we have only one realization $x(.)$ of the time series $X(.)$, then we can consider this realization either in the form

$$x(t) = \sum_{i=1}^{l} y_i f_i(t) + w(t); t = 1, 2, ...,$$

where y_i are realizations of $Y_i; i = 1, 2, ..., l$, or in the form

$$x(t) = \sum_{i=1}^{l} \beta_i f_i(t) + w(t); t = 1, 2, ...,$$

where $\beta_i; i = 1, 2, ..., l$, are some regression parameters. But these two models have different covariance functions. Our aim is now to study the influence of this model choice on predictions, mainly on MSEs of the BLUPs.

In the second model the BLUP $X^*(n + d)$ is given by

$$X^*(n + d) = \hat{X}(n + d) = f'\hat{\beta},$$

where $\hat{\beta} = (F'F)^{-1}F'X$ is the OLSE for β, $f = (f_1(n + d), ..., f_l(n + d))'$ and

$$
\begin{aligned}
MSE_{\sigma^2}[\hat{X}(n + 1)] &= D_{\sigma^2}[X(n + 1)] + \sigma^2 f'(F'F)^{-1}f \\
&= \sigma^2(1 + f'(F'F)^{-1}f).
\end{aligned}
$$

It is clear that $F = (f_1, f_2, ..., f_l)$ and, if $f_i; i = 1, 2, ..., l$, are orthogonal vectors,

$$(F'F)^{-1} = \begin{pmatrix} 1/\|f_1\|^2 & 0 & \cdots & 0 \\ 0 & 1/\|f_2\|^2 & \cdots & 0 \\ \cdot & \cdot & \cdots & \cdot \\ 0 & 0 & \cdots & 1/\|f_l\|^2 \end{pmatrix}$$

and

$$MSE_{\sigma^2}[\hat{X}(n + d)] = \sigma^2 \left(1 + \sum_{i=1}^{l} \frac{1}{\|f_i\|^2} f_i^2(n + d)\right).$$

Comparing the $MSE_\nu[X^*(n + d)]$ and the $MSE_{\sigma^2}[\hat{X}(n + d)]$ we can see that the inequality

$$MSE_\nu[X^*(n + d)] < MSE_{\sigma^2}[\hat{X}(n + d)]$$

is true for every $d > 0$ and for any $\nu \in \Upsilon$.

In a special case when $f_i(t) = \cos \lambda_i t$ or $f_{i+1}(t) = \sin \lambda_i t, \sigma_i^2 = \sigma_{i+1}^2; i = 1, 2, ..., l/2$, and λ_i are Fourier frequencies we have $\|f_i\|^2 = 2/n$ and

$$
\begin{aligned}
MSE_\nu[X^*(n + d)] &= \sigma^2 \left(1 + \sum_{i=1}^{l} \frac{1}{\sigma^2/\sigma_i^2 + n/2} f_i^2(n + d)\right) \\
&= \sigma^2 \left(1 + \sum_{i=1}^{\frac{l}{2}} \frac{2\sigma_i^2}{2\sigma^2 + n\sigma_i^2}\right),
\end{aligned}
$$

since $\cos^2 \lambda_i(n+d) + \sin^2 \lambda_i(n+d) = 1$. We remark that the $MSE_\nu[X^*(n+d)]$ does not depend on d.

For the second model we get

$$(F'F)^{-1} = \begin{pmatrix} 2/n & 0 & \cdots & 0 \\ 0 & 2/n & \cdots & 0 \\ \cdot & \cdot & \cdots & \cdot \\ 0 & 0 & \cdots & 2/n \end{pmatrix}$$

and

$$\begin{aligned} MSE_{\sigma^2}[\hat{X}(n+1)] &= \sigma^2[1 + f'(F'F)^{-1}f] \\ &= \sigma^2\left(1 + \frac{l}{n}\right), \end{aligned}$$

since, under these assumptions, we have

$$f'(F'F)^{-1}f = \frac{2}{n}\sum_{i=1}^{\frac{l}{2}}[\cos^2 \lambda_i(n+1) + \sin^2 \lambda_i(n+1)] = \frac{l}{n}.$$

In the following examples we give some numerical results to illustrate the given results.

Example 4.3.3. Let us consider a time series $X(.)$ given by the model

$$\begin{aligned} X(t) &= Y_1 \cos \lambda_1 t + Y_2 \sin \lambda_1 t + Y_3 \cos \lambda_2 t + Y_4 \sin \lambda_2 t + w(t); t = 1, 2, ..., \\ D[Y_1] &= D[Y_2] = \sigma_1^2, D[Y_3] = D[Y_4] = \sigma_2^2, Cov(Y_i; Y_j) = 0 \text{ for } i \neq j, \end{aligned}$$

where $w(.)$ is a white noise uncorrelated with $Y = (Y_1, Y_2, ..., Y_l)'$ and with variance $D[w(t)] = \sigma^2$,

We know that $X(.)$ is covariance stationary with

$$R_\nu(t) = \sigma_1^2 \cos \lambda_1 t + \sigma_2^2 \cos \lambda_2 t + \sigma^2 \delta_t; t = 1, 2, ...,$$

where

$$\nu = (\sigma^2, \sigma_1^2, \sigma_2^2)' \in \Upsilon = (0, \infty)^3.$$

If $\lambda_1 = \pi/3, \lambda_2 = 2\pi/3$, are Fourier frequencies, $\sigma_1^2 = 2, \sigma_2^2 = 0.1, \sigma^2 = 1$, and $n = 6$, then $D[X(t)] = 3.1, r_\nu'\Sigma_\nu^{-1}r_\nu = 1.74$, and

$$\begin{aligned} MSE_\nu[X^*(n+1)] &= D_\nu[X(n+1)] - r_\nu'\Sigma_\nu^{-1}r_\nu \\ &= 1.36. \end{aligned}$$

If we use for $X(.)$ the LRM

$$X(t) = \beta_1 \cos \lambda_1 t + \beta_2 \sin \lambda_1 t + \beta_3 \cos \lambda_2 t + \beta_4 \sin \lambda_2 t + w(t),$$

then we get, for $\sigma^2 = 1$ and $n = 6$,

$$
\begin{aligned}
MSE_{\sigma^2}[\hat{X}(n+1)] &= \sigma^2 \left(1 + \frac{l}{n}\right) \\
&= 1.66.
\end{aligned}
$$

The difference between these two MSEs is not too large. We see that the first model is better than the second for the given value of parameter ν as it should be.

Example 4.3.4. Let us consider the model

$$
\begin{aligned}
X(t) &= X \cos \lambda_1 t + w(t); E[X] = 0, \\
D[X] &= \sigma_1^2, D[w(t)] = \sigma^2; \nu = (\sigma_1^2, \sigma^2)
\end{aligned}
$$

and the model

$$X(t) = \beta \cos \lambda_1 t + w(t), D[w(t)] = \sigma^2.$$

We remark that in this example the first model is not covariance stationary, since

$$R_\nu(s,t) = \sigma_1^2 \cos \lambda_1 s \cos \lambda_1 t + \sigma^2 \delta_{st}; s, t = 1, 2, ..., n.$$

Let $n = 6, \lambda_1 = \pi/3, \nu = (\sigma_1^2, \sigma^2)', \sigma_1^2 = 10$, and $\sigma^2 = 0.1$. Then we get

$$
\begin{aligned}
MSE_\nu[X^*(7)] &= 0.1 \left(1 + \frac{20}{0.2 + 60} 0.25\right) \\
&= 0.10831.
\end{aligned}
$$

In the second model we get

$$
\begin{aligned}
MSE_{\sigma^2}[\hat{X}(7)] &= D_{\sigma^2}[X(n+1)] + f' Cov_{\sigma^2}(\hat{\beta}) f \\
&= 0.1[1 + 0.25\frac{2}{6}] \\
&= 0.10833
\end{aligned}
$$

and we can see that the MSEs are practically the same.

Example 4.3.5. Let us consider an LRM with correlated errors

$$
\begin{aligned}
X(t) &= \beta_1 \cos \lambda_1 t + \beta_2 \sin \lambda_1 t + X_2 \cos \lambda_2 t + Y_2 \sin \lambda_2 t + w(t); \\
D[X_2] &= D[Y_2] = \sigma_2^2 \text{ and } D[w(t)] = \sigma^2, \nu = (\sigma_2^2, \sigma^2),
\end{aligned}
$$

when

$$
R_\nu(t) = \sigma_2^2 \cos \lambda_2 t + \sigma^2 \delta_t; t = 0, 1, 2, \ldots.
$$

In this model we have

$$
MSE_\nu[X^*(n+1)] = D_\nu[X(n+1)] - r_\nu' \Sigma_\nu^{-1} r_\nu + \left\| f - F' \Sigma_\nu^{-1} r_\nu \right\|_{Cov_\nu(\beta^*)}^2.
$$

Let $n = 6$, $\lambda_1 = \pi/3$, $\lambda_2 = 2\pi/3$, $(\lambda_1, \lambda_2$ are Fourier frequencies), $\sigma_2^2 = 10$, and $\sigma^2 = 1$. Then, after some algebra, we get

$$
MSE_\nu[X^*(7)] = 1.65.
$$

We compare this model with the model

$$
X(t) = \beta_1 \cos \lambda_1 t + \beta_2 \sin \lambda_1 t + \beta_3 \cos \lambda_2 t + \beta_4 \sin \lambda_2 t + w(t),
$$

for which

$$
R_{\sigma^2}(t) = \sigma^2 \delta_t; t = 0, 1, 2, \ldots.
$$

For this model we have, for $\sigma^2 = 1$,

$$
MSE_{\sigma^2}[X^*(7)] = 1.66.
$$

We see that the MSEs are again practically equivalent.

The following lemma gives a basic result for a model with two random components.

Lemma 4.3.2. *For any $n \times 1$ vectors f_1, f_2 and any real positive numbers σ^2, σ_1^2 and σ_2^2 we have*

$$
(\sigma^2 I + \sigma_1^2 f_1 f_1' + \sigma_2^2 f_2 f_2')^{-1} = \frac{1}{\sigma^2} \left(I - \frac{d_1 V_1 + d_2 V_2 - d_1 d_2 (f_1, f_2) V_{12}}{1 - d_1 d_2 (f_1, f_2)^2} \right),
$$

where

$$
d_i = \frac{\sigma_i^2/\sigma^2}{1 + (\sigma_i^2/\sigma^2) \|f_i\|^2}, V_i = f_i f_i'; i = 1, 2, \text{ and } V_{12} = f_1 f_2' + f_2 f_1'.
$$

Proof. From Lemma 4.3.1 we get

$$
(A + \sigma_2^2 f_2 f_2')^{-1} = A^{-1} - \frac{\sigma_2^2 A^{-1} f_2 f_2' A^{-1}}{1 + \sigma_2^2 f_2' A^{-1} f_2}
$$

for any positive definite A and using the expression

$$A^{-1} = (\sigma^2 I + \sigma_1^2 f_1 f_1')^{-1} = \frac{1}{\sigma^2} (I - d_1 V_1)$$

we can write

$$
\begin{aligned}
\sigma^2 (\sigma^2 I + \sigma_1^2 f_1 f_1' + \sigma_2^2 f_2 f_2')^{-1} &= I - d_1 V_1 \\
&\quad - \frac{(\sigma_2^2/\sigma^2)\,(I - d_1 V_1)\,f_2 f_2'\,(I - d_1 V_1)}{1 + (\sigma_2^2/\sigma^2)\,f_2'\,(I - d_1 V_1)\,f_2} \\
&= I - d_1 V_1 \\
&\quad - \frac{(\sigma_2^2/\sigma^2)\,f_2 f_2'}{1 + (\sigma_2^2/\sigma^2)\,(f_2' f_2 - d_1 f_2' f_1 f_1' f_2)} \\
&\quad + \frac{d_1\,(f_1 f_1' f_2 f_2' + f_2 f_2' f_1 f_1')}{1 + (\sigma_2^2/\sigma^2)\,(f_2' f_2 - d_1 f_2' f_1 f_1' f_2)} \\
&\quad - \frac{(\sigma_2^2/\sigma^2)\,d_1^2 f_1 f_1' f_2 f_2' f_1 f_1'}{1 + (\sigma_2^2/\sigma^2)\,(f_2' f_2 - d_1 f_2' f_1 f_1' f_2)} \\
&= I - d_1 V_1 \\
&\quad - \frac{(\sigma_2^2/\sigma^2)\,(V_2 - d_1(f_1, f_2) V_{12})}{1 + (\sigma_2^2/\sigma^2)\left(\|f_2\|^2 - d_1(f_1, f_2)^2 \right)} \\
&\quad - \frac{(\sigma_2^2/\sigma^2)\,d_1^2(f_1, f_2)^2 V_1}{1 + (\sigma_2^2/\sigma^2)\left(\|f_2\|^2 - d_1(f_1, f_2)^2 \right)}.
\end{aligned}
$$

After some algebra we get

$$
\begin{aligned}
1 + \frac{\sigma_2^2}{\sigma^2}\left(\|f_2\|^2 - d_1(f_1, f_2)^2 \right) &= 1 + \frac{\sigma_2^2}{\sigma^2} \|f_2\|^2 \\
&\quad - \frac{\sigma_2^2}{\sigma^2} \frac{(\sigma_1^2/\sigma^2)}{1 + (\sigma_1^2/\sigma^2)\,\|f_1\|^2}(f_1, f_2)^2 \\
&= 1 + \frac{\sigma_2^2}{\sigma^2} \|f_2\|^2 - d_1 \frac{\sigma_2^2}{\sigma^2}(f_1, f_2)^2 \\
&= \left(1 + \frac{\sigma_2^2}{\sigma^2} \|f_2\|^2 \right)(1 - d_1 d_2 (f_1, f_2)^2)
\end{aligned}
$$

and thus

$$
\begin{aligned}
\frac{(\sigma_2^2/\sigma^2)\,(V_2 - d_1(f_1, f_2) V_{12} + d_1^2(f_1, f_2)^2 V_1)}{1 + (\sigma_2^2/\sigma^2)\left(\|f_2\|^2 - d_1(f_1, f_2)^2 \right)} &= \frac{d_2 V_2 - d_1 d_2 (f_1, f_2) V_{12}}{1 - d_1 d_2 (f_1, f_2)^2} \\
&\quad + \frac{d_1^2 d_2 (f_1, f_2)^2 V_1}{1 - d_1 d_2 (f_1, f_2)^2}
\end{aligned}
$$

Using this result we get

$$\sigma^2(\sigma^2 I + \sigma_1^2 f_1 f_1' + \sigma_2^2 f_2 f_2')^{-1} = I - d_1 V_1$$
$$- \frac{d_2 V_2 - d_1 d_2 (f_1, f_2) V_{12}}{1 - d_1 d_2 (f_1, f_2)^2}$$
$$- \frac{d_1^2 d_2 (f_1, f_2)^2 V_1}{1 - d_1 d_2 (f_1, f_2)^2}$$
$$= I - \frac{d_1 V_1 + d_2 V_2 - d_1 d_2 (f_1, f_2) V_{12}}{1 - d_1 d_2 (f_1, f_2)^2}$$

and the lemma is proved.

This result can be used to compare the MSE of the BLUP in the model

$$X(t) = Y_1 f_1(t) + Y_2 f_2(t) + w(t); t = 1, 2, ...,$$
$$D[Y_1] = \sigma_1^2, D[Y_2] = \sigma_2^2,$$
$$D[w(T)] = \sigma^2; \nu = (\sigma_1^2, \sigma_2^2, \sigma^2)',$$

with the MSE of the BLUP in the model

$$X(t) = \beta_1 f_1(t) + \beta_2 f_2(t) + w(t); t = 1, 2, ..., D[w(T)] = \sigma^2,$$

where the $n \times 1$ vectors f_1 and f_2 are not orthogonal.

In the first model we have

$$MSE_\nu[X^*(n + d)] = D_\nu[X(n + d)] - r_\nu' \Sigma_\nu^{-1} r_\nu,$$

where $\Sigma_\nu = \sigma^2 I + \sigma_1^2 f_1 f_1' + \sigma_2^2 f_2 f_2'$, $r_\nu = \sigma_1^2 f_1(n + d) f_1 + \sigma_2^2 f_2(n + d) f_2$, and where we can use Lemma 4.3.2 to compute the Σ_ν^{-1}. The expression for $r_\nu' \Sigma_\nu^{-1} r_\nu$ consists of too many components and we shall not give it here.

In the second model we have

$$MSE_{\sigma^2}[X^*(n + d)] = MSE_{\sigma^2}[\hat{X}(n + d)] = \sigma^2[1 + f'(F'F)^{-1} f],$$

where $f = (f_1(n + d), f_2(n + d))'$ and

$$(F'F)^{-1} = \frac{1}{\|f_1\|^2 \|f_2\|^2 \left(1 - (f_1, f_2)^2 / \left(\|f_1\|^2 \|f_2\|^2\right)\right)} \begin{pmatrix} \|f_2\|^2 & -(f_1, f_2) \\ -(f_1, f_2) & \|f_1\|^2 \end{pmatrix}.$$

From this we get, after some algebra,

$$MSE_{\sigma^2}[\hat{X}(n + d)] = \sigma^2 \left(1 + \frac{f_1^2(n + d)/\|f_1\|^2 + f_2^2(n + d)/\|f_2\|^2}{1 - (f_1, f_2)^2 / \left(\|f_1\|^2 \|f_2\|^2\right)}\right)$$
$$- 2\sigma^2 \frac{f_1(n + d) f_2(n + d)/\left(\|f_1\|^2 \|f_2\|^2\right)}{1 - (f_1, f_2)^2 / \left(\|f_1\|^2 \|f_2\|^2\right)} (f_1, f_2).$$

These results will be used in the following example.

Example 4.3.6. Let us consider the *random linear trend model (RLTM):*

$$
\begin{aligned}
X(t) &= Y_1 + Y_2 t + w(t); t = 1, 2, \ldots, \\
D[Y_1] &= \sigma_1^2, D[Y_2] = \sigma_2^2, \\
D[w(T)] &= \sigma^2; \nu = (\sigma_1^2, \sigma_2^2, \sigma^2)',
\end{aligned}
$$

in which $f_1(t) = 1$ and $f_2(t) = t$, f_1, f_2 *are not orthogonal* and

$$
R_\nu(s, t) = \sigma_1^2 + \sigma_2^2 st + \sigma^2 \delta_{st}.
$$

We shall compare this model with the *random intercept model (RIM):*

$$
\begin{aligned}
X(t) &= \beta + Y_2 t + w(t); t = 1, 2, \ldots, \beta \in E^1, \\
D[Y_2] &= \sigma_2^2, D[w(T)] = \sigma^2; \nu = (\sigma_2^2, \sigma^2)',
\end{aligned}
$$

where

$$
R_\nu(s, t) = \sigma_2^2 st + \sigma^2 \delta_{st}
$$

and with the *random slope model (RSM)*

$$
\begin{aligned}
X(t) &= Y_1 + \beta t + w(t); t = 1, 2, \ldots, \\
D[Y_1] &= \sigma_1^2, D[w(T)] = \sigma^2; \nu = (\sigma_1^2, \sigma^2)',
\end{aligned}
$$

with

$$
R_\nu(s, t) = \sigma_1^2 + \sigma^2 \delta_{st}.
$$

We shall also consider the *linear trend model (LTM)*

$$
X(t) = \beta_1 + \beta_2 t + w(t); D[w(T)] = \sigma^2.
$$

Let us consider the case when $\sigma_1^2 = 1, \sigma_2^2 = 1, \sigma^2 = 0.1$, and $n = 6$. Then we get, after some algebra,

$$
\begin{aligned}
X^*(7) &= r'_\nu \Sigma_\nu^{-1} X, \\
MSE_\nu[X^*(7)] &= D_\nu[X(7)] - r'_\nu \Sigma_\nu^{-1} r_\nu \\
&= 50.1 - 49.92 = 0.18
\end{aligned}
$$

for the RLTM

$$
\begin{aligned}
X^*(7) &= f' \beta_\nu^* + r'_\nu \Sigma_\nu^{-1}(X - F' \beta_\nu^*), \\
MSE_\nu[X^*(7)] &= D_\nu[X(7)] - r'_\nu \Sigma_\nu^{-1} r_\nu + \| f - F' \Sigma_\nu^{-1} r_\nu \|^2_{Cov_\nu(\beta_\nu^*)} \\
&= 0.23
\end{aligned}
$$

for the RIM

$$
\begin{aligned}
X^*(7) &= f'\beta_\nu^* + r_\nu'\Sigma_\nu^{-1}(X - F'\beta_\nu^*), \\
MSE_\nu[X^*(7)] &= D_\nu[X(7)] - r_\nu'\Sigma_\nu^{-1}r_\nu + \left\| f - F'\Sigma_\nu^{-1}r_\nu \right\|_{Cov_\nu(\beta_\nu^*)}^2 \\
&= 0.18
\end{aligned}
$$

for the RSM, and

$$
\begin{aligned}
X^*(7) &= f'\hat\beta \\
MSE_{\sigma^2}[X^*(7)] &= \sigma^2[1 + f'(F'F)^{-1}f] \\
&= 0.19
\end{aligned}
$$

for the LTM.

We remark that the values of ν, β_ν^*, and f are different in every model. We can conclude that in this example the MSE are practically the same. For $\sigma_1^2 = 1, \sigma_2^2 = 1, \sigma^2 = 1$, and $n = 6$ we get

$$
\begin{aligned}
MSE_\nu[X^*(7)] &= 2.25 \text{ for the RLTM}, \\
&= 1.81 \text{ for the RIM}, \\
&= 1.71 \text{ for the RSM}, \\
&= 1.87 \text{ for the LTM}
\end{aligned}
$$

and we see that in the RLTM the BLUP has the largest MSE which is substantionally larger than the MSEs of the BLUPs in the other models. The differencies betwen others MSEs are very small. The best in this case is again the BLUP in the RSM.

These numerical studies show that in the practical aplications of kriging theory, the choice of the model with random coefficients is equivalent to the choice of the LRM with the same regression functions only in the case when the regressors are orthogonal vectors. In other cases it looks to be better to use the LRM.

4.4   Predictions in Multivariate Models

The aim of this section is to study a method of prediction using multivariate regression models. The method is based on the idea of predicting future values of an observed time series using a multivariate regression model with estimated and predicted regression parameters. It is shown that this approach can be studied by the methods of kriging for finding the BLUPs of future observations. The BLUPs are derived by using the Kronecker product of matrices and their basic properties.

Let us assume, for example, that we observe time series $X_i(t)$, where $t = 1, 2, \ldots, 12;\ i = 1, 2, \ldots, q$, and which consists of monthly observations of some real time series during q years. Mainly in a financial time series of observations of such kind we can see that the structure of observations in each year is similar, but changes from year to year. Thus for modeling the whole observed time series by a regression model it is convenient to use a multivariate model of the form, see Chapter 1,

$$X_i = F\beta_i + \varepsilon_i;\ i = 1, 2, \ldots, q,$$

where we assume that $X_i = (X_i(1), X_i(2), \ldots, X_i(n))';\ i = 1, 2, \ldots, q$, are $n \times 1$ random vectors of observations, F is a known $n \times k$ design matrix, $\beta_i = (\beta_{i1}, \beta_{i2}, \ldots, \beta_{ik})';\ i = 1, 2, \ldots, q$, are unknown $k \times 1$ regression parameters. The random $n \times 1$ error vectors $\varepsilon_i;\ i = 1, 2, \ldots, q$, are assumed to have $E[\varepsilon_i] = 0$ for all $i = 1, 2, \ldots, q$.

The MLRM consists of fitting the q models simultaneously. This can also be done as follows. Let X be the $n \times q$ random matrix, let $X = (X_1 X_2 \ldots X_q), B = (\beta_1 \beta_2 \ldots \beta_q)$ be the $k \times q$ matrix of regression coefficients, and let $\varepsilon = (\varepsilon_1 \varepsilon_2 \ldots \varepsilon_q)$ be the $n \times q$ random matrix of errors. Then we can write the multivariate regression model as

$$X = FB + \varepsilon;\ E[\varepsilon] = 0,$$

or, using the operation Vec, as an univariate one

$$
\begin{aligned}
Vec(X) &= (I_q \otimes F)Vec(B) + Vec(\varepsilon);\ E[Vec(\varepsilon)] = 0,\\
Cov(Vec(X)) &= \Sigma_q \otimes \Sigma,
\end{aligned}
$$

where I_q is the $q \times q$ identity matrix, $Vec(X) = (X_1', X_2', \ldots, X_q')'$ is an $nq \times 1$ random vector, $Vec(B) = (\beta_1', \beta_2', \ldots, \beta_q')'$, $\otimes$ denotes the Kronecker product and Σ_q is a p.d. $q \times q$ covariance matrix. The matrices Σ_q and Σ are assumed to be known in this section. From the last assumption we have $Cov(X_i; X_j) = \Sigma_{q,ij}\Sigma$.

If we assume that the random vectors $X_i;\ i = 1, 2, \ldots, q$, are independent with a common covariance $n \times n$ matrix Σ, then $Cov(Vec(X)) = I_q \otimes \Sigma$.

It was shown in Section 1.3 that the BLUE $Vec(B)^*$ of $Vec(B)$ is $Vec(B)^* = (\beta_1^{*'}, \beta_2^{*'}, \ldots, \beta_q^{*'})'$, where

$$\beta_i^* = (F'\Sigma^{-1}F)^{-1}F'\Sigma^{-1}X_i;\ i = 1, 2, \ldots, q$$

for any Σ_q, and thus the OLSE for $Vec(B)$ is $(\hat{\beta}_1', \hat{\beta}_2', \ldots, \hat{\beta}_q')'$, where

$$\hat{\beta}_i = (F'F)^{-1}F'X_i;\ i = 1, 2, \ldots, q.$$

From the point of view of prediction it is reasonable to use a growth curve model instead of the original multivariate regression model. The reason for this change is the following one.

Let us assume that $Vec(X) = (X_1', X_2', \ldots, X_q')'$ is the observed time series modeled by the multivariate regression model and let

$$X_{q+1} = (X_{q+1}(1), X_{q+1}(2), \ldots, X_{q+1}(n))'$$

be the future observation (in the $(q+1)$th year) which should be predicted. We shall assume that

$$E[X_{q+1}] = F\beta_{q+1}.$$

This means that if U is some $X_{q+1}(t); t = 1, 2, \ldots, n$, then we assume that $E[U] = f'\beta_{q+1}$, where $f = (f_1, f_2, \ldots, f_k)'$ is the tth row of F.

Now let $\tilde{\beta}_1, \ldots, \tilde{\beta}_q$ be some estimators of $\beta_1, \ldots, \beta_q$. As a special case we can take $\tilde{\beta}_i = \hat{\beta}_i; i = 1, 2, \ldots, q$.

Let

$$\tilde{\beta}_i = (\tilde{\beta}_{i1}, \tilde{\beta}_{i2}, \ldots, \tilde{\beta}_{ik})'; i = 1, 2, \ldots, q,$$

and let

$$\tilde{\beta}_{.j} = (\tilde{\beta}_{1j}, \tilde{\beta}_{2j}, \ldots, \tilde{\beta}_{qj})'$$

be the $q \times 1$ vector of estimators of the jth component of $\beta_1, \ldots, \beta_q$ for $j = 1, 2, \ldots, k$.

Let us assume that

$$E[\tilde{\beta}_{.j}] = H\gamma_j; j = 1, 2, \ldots, k,$$

where H is a known $q \times l$ matrix, $\gamma_j = (\gamma_{j1}, \gamma_{j2}, \ldots, \gamma_{jl})'; j = 1, 2, \ldots, k$, are unknown $l \times 1$ regression parameters.

If $\tilde{\beta}_1, \ldots, \tilde{\beta}_q$ are unbiased estimators of $\beta_1, \ldots, \beta_q$, then $E[\tilde{\beta}_{.j}] = \beta'_{.j} = \gamma'_j H'$ and, since $\beta'_{.j}$ is the jth row of B, we can write

$$B = \Gamma H',$$

where

$$\Gamma = (\gamma_1 \gamma_2 \ldots \gamma_k)'$$

is the $k \times l$ matrix of the new regression parameters. Thus from the MLRM for X we get the *multivariate growth curve model*

$$X = F\Gamma H' + \varepsilon; E[\varepsilon] = 0,$$

where Γ is a matrix of the new regression parameters. Another form of this model is

$$Vec(X) = (H \otimes F)Vec(\Gamma) + Vec(\varepsilon); E[Vec(\varepsilon)] = 0, Cov(Vec(X)) = \Sigma_q \otimes \Sigma.$$

Define $\tilde{\beta}_{q+1} = (\tilde{\beta}_{q+1,1}, \tilde{\beta}_{q+1,2}, ..., \tilde{\beta}_{q+1,k})'$ by

$$\tilde{\beta}_{q+1,j} = h'\tilde{\gamma}_j; j = 1, 2, \ldots, k,$$

where $h = (h_1, \ldots, h_l)'$ is a given $l \times 1$ vector and $\tilde{\gamma}_j; j = 1, 2, \ldots, k$, are some estimators of $\gamma_j; j = 1, 2, \ldots, k$. Then we get

$$\tilde{\beta}_{q+1} = \tilde{\Gamma} h,$$

where $\tilde{\Gamma} = (\tilde{\gamma}_1 \tilde{\gamma}_2 ... \tilde{\gamma}_k)'$ and, if $\tilde{\gamma}_j$ are unbiased estimators of γ_j for $j = 1, 2, \ldots, k$, we can write

$$E[X_{q+1}] = F\beta_{q+1} = F\Gamma h = (h' \otimes F)Vec(\Gamma)$$

or, if $U = X_{q+1}(t)$ for some t,

$$E[U] = f'\beta_{q+1} = f'\Gamma h = (h' \otimes f')Vec(\Gamma),$$

where f' is the tth row of the matrix F. To make the assumptions clear let us consider the following example.

Example 4.4.1. *Double Linear Trend Prediction.* Let us assume that

$$E[X_i(t)] = \beta_{i1} + \beta_{i2}t; t = 1, 2, ..., n; i = 1, 2, ..., q + 1.$$

Then

$$F = \begin{pmatrix} 1 & 1 & . & . & . & 1 \\ 1 & 2 & . & . & . & n \end{pmatrix}'$$

and for $U = X_{q+1}(t)$ we have $f = (1, t)'; t = 1, 2, ..., n$.

Let $\tilde{\beta}_i = (F'F)^{-1}F'X_i; i = 1, 2, ..., q$, and let $\tilde{\beta}_{.j} = (\tilde{\beta}_{1j}, \tilde{\beta}_{2j}, ..., \tilde{\beta}_{qj})'$, where $j = 1, 2$. If we take $l = 2$,

$$H = \begin{pmatrix} 1 & 1 & . & . & . & 1 \\ 1 & 2 & . & . & . & q \end{pmatrix}'$$

and $h = (1, q + 1)'$, then both the components $\tilde{\beta}_{q+1,j}; j = 1, 2$, of $\tilde{\beta}_{q+1}$ are predicted from $\tilde{\beta}_{.j}; j = 1, 2$, by prolonged regression lines with parameters $\gamma_j; j = 1, 2$, which can also be estimated by the ordinary least squares method by $\tilde{\gamma}_j = (H'H)^{-1}H'\tilde{\beta}_{.j}$ for $j = 1, 2$.

This is an example of a simple approach to prediction in multivariate regression models when the covariance structure of the model given by the covariance matrices Σ_q and Σ is not considered. This approach can also be used in the case when these matrices are unknown.

The kriging approach to prediction in multivariate regression models which we shall now discuss is based on the well-known theory given in the preceding section, and on our assumptions about the matrix X of observations and the random variable U which should be predicted. These can be summarized as follows

$$Vec(X) = (H \otimes F)Vec(\Gamma) + Vec(\varepsilon); E[Vec(\varepsilon)] = 0,$$

and, for $U = X_{q+1}(t)$,

$$E[U] = (h' \otimes f')Vec(\Gamma),$$

where $Vec(\Gamma) = (\gamma_1', \gamma_2', \ldots, \gamma_k')'$ is a new $kl \times 1$ regression parameter and the matrices H, F and vectors h, f are known.

From now on we shall consider the case when $U = X_{q+1}(t)$ for $t = 1, 2, \ldots, n$. Let us assume that the random $n(q+1)$ vector $(Vec(X)', X_{q+1}')'$ has the covariance matrix

$$Cov((Vec(X)', X_{q+1}')') = \begin{pmatrix} \Sigma_q \otimes \Sigma & r_q \otimes \Sigma \\ r_q' \otimes \Sigma & r_{q,q+1}\Sigma \end{pmatrix},$$

where $r_q = (r_{q1}, r_{q2}, \ldots, r_{qq})'$ is a given vector and $r_{q,q+1}$ is a given nonnegative real number. From this assumption we get

$$Cov(X_{q+1}(t); Vec(X)) = r_q \otimes \Sigma_{\bullet t}; t = 1, 2, \ldots, n,$$

where $\Sigma_{\bullet t}$ denotes the tth column of Σ. We shall assume that this $qn \times 1$ vector of covariances is known.

It is well known that in the above prediction regression model the BLUP, $X_{q+1}^*(t)$, of $X_{q+1}(t)$ is given by

$$
\begin{aligned}
X_{q+1}^*(t) &= (h' \otimes f')Vec(\Gamma)^* \\
&\quad + (r_q \otimes \Sigma_{.t})'(\Sigma_q \otimes \Sigma)^{-1}[Vec(X) - (H \otimes F)Vec(\Gamma)^*] \\
&= (h' \otimes f')Vec(\Gamma)^* + ((r_q'\Sigma_q^{-1}) \otimes e_t') \\
&\quad \times [Vec(X) - (H \otimes F)Vec(\Gamma)^*] \\
&= (h' \otimes f')Vec(\Gamma)^* + ((r_q'\Sigma_q^{-1}) \otimes e_t')Vec(X - F\Gamma^*H') \\
&= (h' \otimes f')Vec(\Gamma)^* + e_t'[X - F\Gamma^*H']\Sigma_q^{-1}r_q \\
&= f'\Gamma^*h + r_q'\Sigma_q^{-1}[X - F\Gamma^*H']'e_t,
\end{aligned}
$$

where e_t is the $n \times 1$ vector $(0, 0, \ldots, 0, 1, 0, \ldots, 0)'$ with 1 in the tth place, and $Vec(\Gamma)^*$ is the BLUE of $Vec(\Gamma)$. This is given by

$$
\begin{aligned}
Vec(\Gamma)^* &= ((H \otimes F)'(\Sigma_q \otimes \Sigma)^{-1}(H \otimes F))^{-1} \\
&\quad \times (H \otimes F)'(\Sigma_q \otimes \Sigma)^{-1}Vec(X) \\
&= (H'\Sigma_q^{-1}H)^{-1}H'\Sigma_q^{-1} \otimes (F'\Sigma^{-1}F)^{-1}F'\Sigma^{-1}Vec(X),
\end{aligned}
$$

from which we get

$$\Gamma^* = (F'\Sigma^{-1}F)^{-1}F'\Sigma^{-1}X\Sigma_q^{-1}H(H'\Sigma_q^{-1}H)^{-1}.$$

The last expression can also be written in the form

$$\begin{aligned} \Gamma^* &= (\gamma_1^*\gamma_2^*...\gamma_k^*)' \\ &= B^*\Sigma_q^{-1}H(H'\Sigma_q^{-1}H)^{-1} \\ &= (\beta_1^*\beta_2^* \ldots \beta_q^*)\Sigma_q^{-1}H(H'\Sigma_q^{-1}H)^{-1}, \end{aligned}$$

and we get, for the BLUEs γ_j^*; $j = 1, 2, \ldots, k$, the expressions

$$\gamma_j^* = (H'\Sigma_q^{-1}H)^{-1}H'\Sigma_q^{-1}\beta_{.j}^*; j = 1, 2, \ldots, k.$$

It is also easy to derive the expression for the covariance matrix of the estimator $Vec(\Gamma)^*$. This is given by

$$Cov(Vec(\Gamma)^*) = (H'\Sigma_q^{-1}H)^{-1} \otimes (F'\Sigma^{-1}F)^{-1}.$$

Since $\Gamma^*h = \beta_{q+1}^*$ and $\Gamma^*H' = B^*$ we can also write the BLUP $X_{q+1}^*(t)$ in the form

$$\begin{aligned} X_{q+1}^*(t) &= f'\beta_{q+1}^* + r_q'\Sigma_q^{-1}[X - FB^*]'e_t \\ &= f'\beta_{q+1}^* + r_q'\Sigma_q^{-1}[X(t) - (F\beta^*)_t]_q, \end{aligned}$$

where $[X(t) - (F\beta^*)_t]_q$ denotes the $q \times 1$ random vector with components $X_j(t) - (F\beta_j^*)_t; j = 1, 2, ..., q$.

Now we shall derive the MSE of the predictor $X_{q+1}^*(t)$. We have

$$MSE[X_{q+1}^*(t)] = D[X_{q+1}(t)] - (r_q \otimes \Sigma_{\bullet t})'(\Sigma_q \otimes \Sigma)^{-1}(r_q \otimes \Sigma_{\bullet t}) + a,$$

where

$$a = \left\| h \otimes f - (H' \otimes F')(\Sigma_q \otimes \Sigma)^{-1}(r_q \otimes \Sigma_{\bullet t}) \right\|_{Cov(Vec(\Gamma)^*)}^2.$$

Using the expression

$$(r_q \otimes \Sigma_{\bullet t})'(\Sigma_q \otimes \Sigma)^{-1}(r_q \otimes \Sigma_{\bullet t}) = r_q'\Sigma_q^{-1}r_q \Sigma_{tt}$$

and

$$\begin{aligned} h \otimes f - (H' \otimes F')(\Sigma_q \otimes \Sigma)^{-1}(r_q \otimes \Sigma_{.t}) &= h \otimes f \\ &\quad -H'\Sigma_q^{-1}r_q \otimes F' \Sigma^{-1}\Sigma_{.t} \\ &= (h - H'\Sigma_q^{-1}r_q) \otimes f \end{aligned}$$

we get the following expression for the MSE of the predictor $X_{q+1}^*(t)$:

$$
\begin{aligned}
E[X_{q+1}^*(t) - X_{q+1}(t)]^2 &= D[X_{q+1}(t)] - r_q' \Sigma_q^{-1} r_q \Sigma_{tt} \\
&\quad + (h - H'\Sigma_q^{-1} r_q)'(H'\Sigma_q^{-1} H)^{-1}(h - H'\Sigma_q^{-1} r_q) \\
&\quad f'(F'\Sigma^{-1} F)^{-1} f.
\end{aligned}
$$

This expression can also be written in the form

$$
\begin{aligned}
E[X_{q+1}^*(t) - X_{q+1}(t)]^2 &= (r_{q,q+1} - r_q' \Sigma_q^{-1} r_q) \Sigma_{tt} \\
&\quad + (h - H'\Sigma_q^{-1} r_q)'(H'\Sigma_q^{-1} H)^{-1}(h - H'\Sigma_q^{-1} r_q) \\
&\quad f'(F'\Sigma^{-1} F)^{-1} f.
\end{aligned}
$$

It can easily be seen that in the case when $\Sigma_q = \sigma_1^2 I, \Sigma = \sigma^2 I_n$, and $r_q = 0$ we get the BLUP, $X_{q+1}^*(t) = \hat{X}_{q+1}(t)$, in the form

$$
\hat{X}_{q+1}(t) = f'\hat{\beta}_{q+1},
$$

where $\hat{\beta}_{q+1} = \hat{\Gamma} h$ with $\hat{\Gamma} = (\hat{\gamma}_1 \hat{\gamma}_2 ... \hat{\gamma}_k)'$, $\hat{\gamma}_j = (H'H)^{-1} H' \hat{\beta}_{.j}$, and $\hat{\beta}_j$ are the OLSEs given by $\hat{\beta}_j(X) = (F'F)^{-1} F' X_j; j = 1, 2, ..., k$.

The MSE of this predictor is given by

$$
E[\hat{X}_{q+1}(t) - X_{q+1}(t)]^2 = \sigma_1^2 \sigma^2 [1 + h'(H'H)^{-1} h \times f'(F'F)^{-1} f].
$$

The BLUP $X_{q+1}^*(t)$ was derived under the condition that the covariance matrix $Cov((Vec(X)', X_{q+1}')')$ is known. This is a rather strong condition which is usually not fulfilled in practical problems of the prediction of time series. The other thing is that the prediction regression model for X and X_{q+1}, with unknown covariance matrix $Cov((Vec(X)', X_{q+1}')')$, has too many parameters which should be estimated if they are unknown. The number of parameters can be reduced using some parametric models for this covariance matrix. One simple parametrization is given in the following example. The problem is similar to that studied in Fuller and Hasza (1981) and Štulajter (1994b).

Example 4.4.2. Let us assume that

$$
\Sigma_{q,ij} = \frac{\sigma_1^2}{1 - \rho_1} \rho_1^{|i-j|}; i, j = 1, 2, ..., q,
$$

where $\sigma_1^2 \geq 0$ and $|\rho_1| < 1$ and $r_q = [\sigma_1^2/(1 - \rho_1^2)](\rho_1^q, \rho_1^{q-1}, ..., \rho_1)'$. Then we get, using the equality $r_q' \Sigma_q^{-1} = (0, 0, ..., 0, \rho_1)'$:

$$
X_{q+1}^*(t) = \sum_{j=1}^k f_j h'(H'\Sigma_q^{-1} H)^{-1} H'\Sigma_q^{-1} \beta_{.j}^* + \rho_1(X_q(t) - (F\beta_q^*)_t).
$$

Taking $H = (1, 1, \ldots, 1)'$ and $h = 1$ we have, after some algebra,

$$X_{q+1}^*(t) = \sum_{j=1}^{k} f_j \frac{\beta_{1j}^* + \beta_{qj}^* + (1 - \rho_1) \sum_{i=2}^{q-1} \beta_{ij}^*}{2 + (q-2)(1 - \rho_1)} + \rho_1(X_q(t) - (F\beta_q^*)_t).$$

This predictor has the MSE

$$E[X_{q+1}^*(t) - X_{q+1}(t)]^2 = \sigma_1^2 \Sigma_{tt} + \frac{1 - \rho_1}{2 + (q-2)(1 - \rho_1)} f'(F'\Sigma^{-1}F)^{-1} f.$$

In the special case when also $\Sigma_{ij} = \left[\sigma^2 / \left(1 - \rho^2\right)\right] \rho^{|i-j|}; i, j = 1, 2, \ldots, n$ for some $\sigma^2 \geq 0$ and $\rho < 1$ and $F = (1, 1, \ldots, 1)', f = 1$, we have

$$\beta_i^* = \frac{X_i(1) + X_i(n) + (1 - \rho) \sum_{t=2}^{n-1} X_i(t)}{2 + (n-2)(1 - \rho)}; i = 1, 2, \ldots, q,$$

$$f'(F'\Sigma^{-1}F)^{-1} f = \frac{\sigma^2}{(1 - \rho)[2 + (n-2)(1 - \rho)]},$$

and $\Sigma_{tt} = \sigma^2 / \left(1 - \rho^2\right)$.

From the practical point of view it is possible to use the last predictor with estimated parameters ρ and ρ_1 (and with estimated σ^2 and σ_1^2 to find an estimator of its MSE).

Now we shall study problems of predictions in RLRMs with the correlated components. Let us assume that we again observe the time series $X_i(t); t = 1, 2, \ldots, 12; i = 1, 2, \ldots, q$, which consists of monthly observations of some real time series during q years. Now we shall assume that the mean values of $X_i(.); i = 1, 2, \ldots, q$ are given by the same LRM. Thus to model the whole observed time series by a regression model it is convenient to use an RLRM of the form, see Chapter 1,

$$X_i = F\beta + \varepsilon_i; i = 1, 2, \ldots, q,$$

where we assume that $X_i = (X_i(1), X_i(2), \ldots, X_i(n))'; i = 1, 2, \ldots, q$ are $n \times 1$ random vectors of observations, F is a known $n \times k$ design matrix, and $\beta = (\beta_1, \beta_2, \ldots, \beta_k)'$ is an unknown $k \times 1$ regression parameter. The random $n \times 1$ error vectors $\varepsilon_i; i = 1, 2, \ldots, q$, are assumed to have $E[\varepsilon_i] = 0$ for all $i = 1, 2, \ldots, q$. Let X be an $n \times p$ random matrix given by $X = (X_1 \ldots X_p)$. Then the RLRM can be written as

$$X = F\beta j_q' + \varepsilon; E[\varepsilon] = 0,$$

where j_q is the $q \times 1$ vector $j_q = (1, 1, \ldots, 1)'$ and $\varepsilon = (\varepsilon_1 \ldots \varepsilon_q)$ is an $n \times q$ matrix of random errors.

Using the operation Vec we can write a RLRM with correlated components as

$$Vec(X) = (j_q \otimes F)\beta + Vec(\varepsilon), E[Vec(\varepsilon)] = 0, Cov(Vec(X)) = \Sigma_q \otimes \Sigma,$$

where we assume that Σ_q and Σ are known covariance matrices. Let

$$X_{q+1} = (X_{q+1}(1), X_{q+1}(2), \ldots, X_{q+1}(n))'$$

be the future observation (in the $(q+1)$th year) which should be predicted. We shall assume that

$$E[X_{q+1}] = F\beta$$

and, as before, that the random $n(q+1)$ vector $(Vec(X)', X'_{q+1})'$, has the known covariance matrix

$$Cov((Vec(X)', X'_{q+1})') = \begin{pmatrix} \Sigma_q \otimes \Sigma & r_q \otimes \Sigma \\ r'_q \otimes \Sigma & r_{q,q+1}\Sigma \end{pmatrix},$$

where $r_q = (r_{q1}, r_{q2}, \ldots, r_{qq})'$ is a given vector and $r_{q,q+1}$ is a given nonnegative real number. From this assumption we again get that

$$Cov(X_{q+1}(t); Vec(X)) = r_q \otimes \Sigma_{.t}; t = 1, 2, \ldots, n,$$

where $\Sigma_{.t}$ denotes the tth column of Σ.

Under these assumptions we can derive the following expression for the BLUP, $X^*_{q+1}(t)$, of $X_{q+1}(t)$:

$$X^*_{q+1}(t) = f'\beta^* + (r_q \otimes \Sigma_{.t})'(\Sigma_q \otimes \Sigma)^{-1}[Vec(X) - (j_q \otimes F)\beta^*],$$

where β^* is the BLUE of β, which is now given by

$$
\begin{aligned}
\beta^*(X) &= ((j_q \otimes F)'(\Sigma_q \otimes \Sigma)^{-1}(j_q \otimes F))^{-1}(j_q \otimes F)'(\Sigma_q \otimes \Sigma)^{-1}Vec(X) \\
&= (j'_q\Sigma_q^{-1}j_q \otimes (F'\Sigma^{-1}F))^{-1}(j'_q\Sigma_q^{-1} \otimes F'\Sigma^{-1})Vec(X) \\
&= (j'_q\Sigma_q^{-1}j_q)^{-1}(F'\Sigma^{-1}F)^{-1}\sum_{i=1}^{q}(j'_q\Sigma_q^{-1})_i F'\Sigma^{-1}X_i \\
&= (F'\Sigma^{-1}F)^{-1}F'\Sigma^{-1}\bar{X}_w.
\end{aligned}
$$

$\bar{X}_w$ is the weighted arithmetic mean of $X_i; i = 1, 2, \ldots, q$, given by

$$\bar{X}_w = (j'_q\Sigma_q^{-1}j_q)^{-1}\sum_{i=1}^{q}(j'_q\Sigma_q^{-1})_i X_i.$$

The covariance matrix $Cov(\beta^*)$ of β^* is

$$
\begin{aligned}
Cov(\beta^*) &= (F'\Sigma^{-1}F)^{-1}F'\Sigma^{-1}Cov(\bar{X}_w)\Sigma^{-1}F(F'\Sigma^{-1}F)^{-1} \\
&= (F'\Sigma^{-1}F)^{-1}F'\Sigma^{-1}(j'_q\Sigma_q^{-1}j_q)^{-2} \\
&\quad \times \sum_{i,j=1}^{q}(j'_q\Sigma_q^{-1})_i(j'_q\Sigma_q^{-1})_j Cov(X_i;X_j)\; \Sigma^{-1}F(F'\Sigma^{-1}F)^{-1} \\
&= (F'\Sigma^{-1}F)^{-1}F'\Sigma^{-1}(j'_q\Sigma_q^{-1}j_q)^{-2} \\
&\quad \times \sum_{i,j=1}^{q}(j'_q\Sigma_q^{-1})_i(j'_q\Sigma_q^{-1})_j(\Sigma_{q,ij}\otimes\Sigma)\Sigma^{-1}F(F'\Sigma^{-1}F)^{-1} \\
&= (j'_q\Sigma_q^{-1}j_q)^{-1}(F'\Sigma^{-1}F)^{-1}.
\end{aligned}
$$

After some computation we get, as for the MLRM,

$$
\begin{aligned}
X^*_{q+1}(t) &= f'\beta^*_{q+1} + r'_q\Sigma_q^{-1}[X - F\beta^* j'_q]'e_t \\
&= f'\beta^*_{q+1} + r'_q\Sigma_q^{-1}[X(t) - (F\beta^*)_t]_q,
\end{aligned}
$$

where $[X(t) - (F\beta^*)_t]_q$ denotes the $q \times 1$ random vector with components $X_j(t) - (F\beta^*)_t; j = 1, 2, ..., q$.

The MSE of this predictor $X^*_{q+1}(t)$ is

$$
\begin{aligned}
MSE[X^*_{q+1}(t)] &= D[X_{q+1}(t)] - (r_q\otimes\Sigma_{.t})'(\Sigma_q\otimes\Sigma)^{-1}(r_q\otimes\Sigma_{.t}) \\
&\quad + \left\| f - (j_q\otimes F)'(\Sigma_q\otimes\Sigma)^{-1}(r_q\otimes\Sigma_{.t}) \right\| \\
&= D[X_{q+1}(t)] - r'_q\Sigma_q r_q\Sigma_{tt} \\
&\quad + \left\| f - j'_q\Sigma_q^{-1}r_q\otimes F'\Sigma^{-1}\Sigma_{.t} \right\|^2_{Cov(\beta^*)} \\
&= D[X_{q+1}(t)] - r'_q\Sigma_q r_q\Sigma_{tt} + \left\| f - r'_q\Sigma_q^{-1}j_q\, f \right\|^2_{Cov(\beta^*)} \\
&= D[X_{q+1}(t)] - r'_q\Sigma_q r_q\Sigma_{tt} + (1 - r'_q\Sigma_q^{-1}j_q)^2\, \|f\|^2_{Cov(\beta^*)}.
\end{aligned}
$$

In the special case when $\Sigma_q = I_q$ and $r_q = 0$ we get

$$
\beta^*(X) = (F'\Sigma^{-1}F)^{-1}F'\Sigma^{-1}\bar{X}_w,
$$

where $\bar{X}_w$ is the arithmetic mean of $X_i; i = 1, 2, ..., q$, given by $\bar{X}_w = 1/q\sum_{i=1}^{q}X_i$,

$$
Cov(\beta^*) = \frac{1}{q}(F'\Sigma^{-1}F)^{-1}
$$

and for the BLUP $X^*_{q+1}(t)$ we get

$$X^*_{q+1}(t) = f'\beta^*,$$
$$E[X^*_{q+1}(t) - X_{q+1}(t)]^2 = D[X_{q+1}(t)] + \|f\|^2_{Cov(\beta^*)}.$$

The results for $\Sigma = \sigma^2 I_n$ are obvious.

4.5   Predictions in Nonlinear Models

The kriging method of prediction can also be used for time series whose mean value is given by an NRM . Let

$$X(t) = m_\gamma(t) + \varepsilon(t); t = 1, 2, ...,$$

where the regression parameter $\gamma = (\gamma_1, ..., \gamma_k)' \in \Gamma$ and the time series $\varepsilon(.)$ are assumed to have a mean value equal to zero and a known covariance function $R(s, t); s, t = 1, 2, ...$ Then for any random variable $U = X(n+d)$, where d is a positive integer, we get

$$E_\gamma[U] = m_\gamma(n + d); \gamma \in \Gamma \text{ and } Cov(U; X(t)) = R(n + d, t); t = 1, 2, ...$$

Let $X = (X(1), ..., X(n))'$ be a finite observation of $X(.)$, let U be a predicted random variable, and let $Z = (X', U)'$. Then the random vector Z is given by the *nonlinear prediction regression model*

$$Z = \begin{pmatrix} X \\ U \end{pmatrix} = \begin{pmatrix} m_\gamma \\ m_\gamma(n + d) \end{pmatrix} + \varepsilon; \gamma \in \Gamma;$$
$$E[\varepsilon] = 0; Cov(Z) = \begin{pmatrix} \Sigma & r \\ r' & D[U] \end{pmatrix},$$

where $m_\gamma = (m_\gamma(1), m_\gamma(2), ..., m_\gamma(n))', \Sigma = Cov(X)$ with $\Sigma_{st} = R(s, t)$, and $r = Cov(X, U)$ is an $n \times 1$ vector of covariances between X and U with components $r_t = Cov(X(t); U) = R(t, n + d); t = 1, 2, ..., n$. We shall assume that Σ and r are known. Let $\tilde{\gamma}(X)$ be some estimator of γ based on X. Then we can define the predictor $\tilde{U} = \tilde{X}(n + d)$ by

$$\tilde{U} = m_{\tilde{\gamma}}(n + d) + r'\Sigma^{-1}(X - m_{\tilde{\gamma}}).$$

If we take for $\tilde{\gamma}$ the WELSE $\hat{\gamma}_\Sigma$, defined by

$$\hat{\gamma}_\Sigma(X) = \arg\min_\gamma \|X - m_\gamma\|^2_{\Sigma^{-1}},$$

then we get the predictor $U^* = X^*(n+d)$, the analogue of the BLUP, given by

$$U^* = m_{\hat{\gamma}_\Sigma}(n+d) + r'\Sigma^{-1}(X - m_{\hat{\gamma}_\Sigma}).$$

Denoting by $F_{\gamma_0} = \partial m_\gamma / \partial \gamma' |_{\gamma_0}$ the $n \times k$ matrix with components

$$F_{\gamma_0,ti} = \frac{\partial m_\gamma(t)}{\partial \gamma_i} |_{\gamma = \gamma_0}; t = 1, 2, .., n, i = 1, 2, ..., k,$$

and by f_{γ_0} the $k \times 1$ vector with components

$$f_{\gamma_0,i} = \frac{\partial m_\gamma(n+d)}{\partial \gamma_i} |_{\gamma = \gamma_0}; i = 1, 2, ..., k,$$

we can write, using the Taylor formula near γ_0, the true value of the parameter γ, the approximate expression for $\tilde{U} = \tilde{X}(n+1)$, in the form

$$
\begin{aligned}
\tilde{U} &\approx m_{\gamma_0}(n+1) + f'_{\gamma_0}(\tilde{\gamma} - \gamma_0) + r'\Sigma^{-1}(X - m_{\gamma_0} - F_{\gamma_0}(\tilde{\gamma} - \gamma_0)) \\
&= m_{\gamma_0}(n+1) + r'\Sigma^{-1}(X - m_{\gamma_0}) + (f'_{\gamma_0} - r'\Sigma^{-1}F_{\gamma_0})(\tilde{\gamma} - \gamma_0) \\
&= m_{\gamma_0}(n+1) + r'\Sigma^{-1}(X - m_{\gamma_0}) + (f_{\gamma_0} - F'_{\gamma_0}\Sigma^{-1}r)'(\tilde{\gamma} - \gamma_0) \\
&= U_0^* + (f_{\gamma_0} - F'_{\gamma_0}\Sigma^{-1}r)'(\tilde{\gamma} - \gamma_0),
\end{aligned}
$$

where

$$U_0^* = m_{\gamma_0}(n+1) + r'\Sigma^{-1}(X - m_{\gamma_0})$$

is the BLUP for U by the given value γ_0 of γ.

Next we know that the MSE for any predictor $\tilde{U}$ is given by

$$E_{\gamma_0}[\tilde{U} - U]^2 = D_{\gamma_0}[\tilde{U}] + D_{\gamma_0}[U] + (E_{\gamma_0}[\tilde{U}] - E_{\gamma_0}[U])^2 - 2Cov_{\gamma_0}(\tilde{U}; U).$$

Using the preceding approximation for $\tilde{U}$ we can write

$$
\begin{aligned}
D_{\gamma_0}[\tilde{U}] &\approx D_{\gamma_0}[U_0^*] + D_{\gamma_0}[(f_{\gamma_0} - F'_{\gamma_0}\Sigma^{-1}r)'(\tilde{\gamma} - \gamma_0)] \\
&\quad + 2Cov_{\gamma_0}(U_0^*; (f_{\gamma_0} - F'_{\gamma_0}\Sigma^{-1}r)'(\tilde{\gamma} - \gamma_0)) \\
&= r'\Sigma^{-1}r + (f_{\gamma_0} - F'_{\gamma_0}\Sigma^{-1}r)'Cov_{\gamma_0}(\tilde{\gamma})(f_{\gamma_0} - F'_{\gamma_0}\Sigma^{-1}r) \\
&\quad + 2(f_{\gamma_0} - F'_{\gamma_0}\Sigma^{-1}r)'Cov_{\gamma_0}(\tilde{\gamma}; U_0^*),
\end{aligned}
$$

where

$$Cov_{\gamma_0}(\tilde{\gamma}; U_0^*) = Cov_{\gamma_0}(\tilde{\gamma}; r'\Sigma^{-1}X) = Cov_{\gamma_0}(\tilde{\gamma}; X)\Sigma^{-1}r.$$

Next we have

$$(E_{\gamma_0}[\tilde{U}] - E_{\gamma_0}[U])^2 \approx [(f_{\gamma_0} - F'_{\gamma_0}\Sigma^{-1}r)'(E_{\gamma_0}[\tilde{\gamma}] - \gamma_0)]^2,$$

and

$$
\begin{aligned}
Cov_{\gamma_0}(\tilde{U};U) &\approx Cov_{\gamma_0}(U_0^*;U) + (f_{\gamma_0} - F'_{\gamma_0}\Sigma^{-1}r)'Cov_{\gamma_0}(\tilde{\gamma};U) \\
&= r'\Sigma^{-1}r + (f_{\gamma_0} - F'_{\gamma_0}\Sigma^{-1}r)'Cov_{\gamma_0}(\tilde{\gamma};U).
\end{aligned}
$$

Using these expressions we get

$$
\begin{aligned}
E_{\gamma_0}[\tilde{U} - U]^2 &\approx D_{\gamma_0}[U] - r'\Sigma^{-1}r + \left\| f_{\gamma_0} - F'_{\gamma_0}\Sigma^{-1}r \right\|^2_{Cov_{\gamma_0}(\tilde{\gamma})} \\
&\quad + \left\| f_{\gamma_0} - F'_{\gamma_0}\Sigma^{-1}r \right\|^2_{b_0 b'_0} \\
&\quad + 2(f_{\gamma_0} - F'_{\gamma_0}\Sigma^{-1}r)'(Cov(\tilde{\gamma};X)\Sigma^{-1}r - Cov(\tilde{\gamma};U)) \\
&= D_{\gamma_0}[U] - r'\Sigma^{-1}r + \left\| f_{\gamma_0} - F'_{\gamma_0}\Sigma^{-1}r \right\|^2_{Cov_{\gamma_0}(\tilde{\gamma}) + b_0 b'_0} \\
&\quad + 2(f_{\gamma_0} - F'_{\gamma_0}\Sigma^{-1}r)'(Cov(\tilde{\gamma};X)\Sigma^{-1}r - Cov(\tilde{\gamma};U)),
\end{aligned}
$$

where we have used the notation $b_0(\tilde{\gamma}) = b_0$ for the bias of $\tilde{\gamma}$:

$$b_0(\tilde{\gamma}) = E_{\gamma_0}[\tilde{\gamma}] - \gamma_0.$$

If $\tilde{\gamma}$ is equal to the WELSE $\hat{\gamma}_\Sigma$, then we have an approximation for $\hat{\gamma}_\Sigma$ in terms of ε which can be obtained by using the Taylor expansions for m_γ and F_γ around the true value γ_0 of the parameter γ. We can write, as in Section 3.2,

$$
\begin{aligned}
\hat{\gamma}_\Sigma(\varepsilon) &\approx \gamma_0 + (F'\Sigma^{-1}F)^{-1}F'\Sigma^{-1}\varepsilon \\
&\quad + (F'\Sigma^{-1}F)^{-1}\left[\begin{pmatrix} \varepsilon'N_1\varepsilon \\ \cdot \\ \cdot \\ \varepsilon'N_k\varepsilon \end{pmatrix} - \frac{1}{2}F'\Sigma^{-1}\begin{pmatrix} \varepsilon'A'H_1 A\varepsilon \\ \cdot \\ \cdot \\ \varepsilon'A'H_n A\varepsilon \end{pmatrix} \right] \\
&= \gamma_0 + A\varepsilon + B[\varepsilon'N_\bullet\varepsilon - C\varepsilon'A'H_\bullet A\varepsilon].
\end{aligned}
$$

In these expressions we have simplified the notation by writing

$$F_{\gamma_0} = F, A = (F'\Sigma^{-1}F)^{-1}F'\Sigma^{-1}, B = (F'\Sigma^{-1}F)^{-1},$$

and

$$C = \frac{1}{2}F'\Sigma^{-1}.$$

Next, $\varepsilon' N_\bullet \varepsilon$ denotes the $k \times 1$ random vector with components $\varepsilon' N_j \varepsilon$, where $j = 1, 2, ..., k$ and $\varepsilon' A' H_\bullet A\varepsilon$ denotes the $n \times 1$ random vector with components $\varepsilon' A' H_t A\varepsilon; t = 1, 2, ..., n$.

The $n \times n$ matrices $N_j; j = 1, 2, ..., k$, are defined by $N_j = \frac{1}{2}(O_j + O'_j)$, where

$$O_{j,kl} = \sum_{t=1}^{n} (H_t A)_{jk} M_{\Sigma, tl}; k, l = 1, 2, ..., n,$$

and where $H_t; t = 1, 2, ..., n$ are the $k \times k$ Hessian matrices of m_γ defined by

$$H_{t,ij} = \frac{\partial^2 m_\gamma(t)}{\partial \gamma_i \partial \gamma_j}; i, j = 1, 2, ..., k,$$

and

$$M_\Sigma = I - F(F'\Sigma^{-1}F)^{-1}F'\Sigma^{-1}.$$

Using the expression for a mean value of a quadratic form we get

$$
\begin{aligned}
b_0(\hat{\gamma}_\Sigma) &\approx (F'\Sigma^{-1}F)^{-1} \left[tr(N_\bullet \Sigma) - \frac{1}{2} F'\Sigma^{-1} tr(A' H_\bullet A\Sigma) \right] \\
&= (F'\Sigma^{-1}F)^{-1} \left[tr(N_\bullet \Sigma) - \frac{1}{2} F'\Sigma^{-1} tr(H_\bullet (F'\Sigma^{-1}F)^{-1}) \right],
\end{aligned}
$$

where we have used the notation $tr(N_\bullet \Sigma)$ for the $k \times 1$ vector with components $tr(N_j \Sigma); j = 1, 2, ..., k$, and $tr(A' H_\bullet A\Sigma)$ for the $n \times 1$ vector with components $tr(A' H_t A\Sigma); t = 1, 2, ..., n$, and the equality $A\Sigma A' = (F'\Sigma^{-1}F)^{-1}$.

It should be noted that the approximate bias b_0 depends on the true value γ_0 of the regression parameter γ and also on the covariance matrix Σ of X.

The approximate expression for the covariance matrix $Cov(\hat{\gamma}_\Sigma)$ can also be derived easily. We get, under the assumption that X has normal distribution, that

$$
\begin{aligned}
Cov_{\gamma_0}(\hat{\gamma}_\Sigma) &\approx A\Sigma A' + B[Cov(\varepsilon' N_\bullet \varepsilon - C\varepsilon' A' H_\bullet A\varepsilon)]B' \\
&= A\Sigma A' + B[Cov(\varepsilon' N_\bullet \varepsilon) + Cov(C\varepsilon' A' H_\bullet A\varepsilon)] \\
&\quad -2Cov(C\varepsilon' A' H_\bullet A\varepsilon; \varepsilon' N_\bullet \varepsilon)]B' \\
&= A\Sigma A' + B[2tr(N_\bullet \Sigma N_\bullet \Sigma) + C2tr(A' H_\bullet A\Sigma A' H_\bullet A\Sigma)C' \\
&\quad -2C2tr(A' H_\bullet A\Sigma N_\bullet \Sigma)]B'.
\end{aligned}
$$

Here the notation $tr(N_\bullet \Sigma N_\bullet \Sigma)$ was used for the $k \times k$ matrix with components $tr(N_i \Sigma N_j \Sigma)$, the notation $tr(A' H_\bullet A\Sigma A' H_\bullet A\Sigma)$ for the $n \times n$ matrix with components $tr(A'H_sA\Sigma A'H_tA\Sigma)$, and for the $n \times k$ matrix with components $tr(A'H_tA\Sigma N_j\Sigma)$ the notation $tr(A'H_\bullet A\Sigma N_\bullet\Sigma)$ was used, where $i, j = 1, 2, ..., k$ and $s, t = 1, 2, ..., n$.

Using the expressions for A, B, and C we get

$$
\begin{aligned}
Cov_{\gamma_0}(\hat{\gamma}_\Sigma) &\approx (F'\Sigma^{-1}F)^{-1} + (F'\Sigma^{-1}F)^{-1}[2tr(N_\bullet\Sigma N_\bullet\Sigma) \\
&\quad + \frac{1}{4}F'\Sigma^{-1}2tr(H_\bullet A\Sigma A'H_\bullet A\Sigma A')\Sigma^{-1}F \\
&\quad - F'\Sigma^{-1}2tr(A'H_\bullet A\Sigma N_\bullet\Sigma)](F'\Sigma^{-1}F)^{-1},
\end{aligned}
$$

where

$$A\Sigma A' = (F'\Sigma^{-1}F)^{-1}.$$

It should be remarked that the matrix F, the matrices $N_j; j = 1, 2, ..., k$, and the matrices $H_t; t = 1, 2, ..., n$, in the last expression depend on γ_0.

If we take $\tilde{\gamma}$ equal to the OLSE $\hat{\gamma}$, then we can write

$$
\hat{\gamma}(\varepsilon) \approx \gamma_0 + (F'F)^{-1}F'\varepsilon + (F'F)^{-1}\left[\begin{pmatrix} \varepsilon'N_1\varepsilon \\ \cdot \\ \cdot \\ \cdot \\ \varepsilon'N_k\varepsilon \end{pmatrix} - \frac{1}{2}F'\begin{pmatrix} \varepsilon'A'H_1A\varepsilon \\ \cdot \\ \cdot \\ \cdot \\ \varepsilon'A'H_nA\varepsilon \end{pmatrix}\right],
$$

where now $A = (F'F)^{-1}F'$ and in the definition of matrices $N_j; j = 1, 2, ..., k$, we now use the projection matrix $M = I - F(F'F)^{-1}F'$ instead of the matrix M_Σ.

Thus we get

$$b_0(\hat{\gamma}) \approx (F'F)^{-1}\left[tr(N_\bullet\Sigma) - \frac{1}{2}F'tr(H_\bullet A\Sigma A')\right],$$

where now

$$A\Sigma A' = (F'F)^{-1}F'\Sigma F(F'F)^{-1}.$$

For the covariance matrix $Cov_{\gamma_0}(\hat{\gamma})$ we get

$$
\begin{aligned}
Cov_{\gamma_0}(\hat{\gamma}) &\approx (F'F)^{-1}F'\Sigma F(F'F)^{-1} + (F'F)^{-1}[2tr(N_\bullet\Sigma N_\bullet\Sigma) \\
&\quad + \frac{1}{4}F'2tr(H_\bullet A\Sigma A'H_\bullet A\Sigma A')F \\
&\quad - F'2tr(A'H_\bullet A\Sigma N_\bullet\Sigma)](F'F)^{-1}.
\end{aligned}
$$

Again the matrix F, the matrices $N_j; j = 1, 2, ..., k$, and the matrices $H_t; t = 1, 2, ..., n$, in the last expression depend on γ_0.

For $\tilde{\gamma}(\varepsilon) = \hat{\gamma}_\Sigma(\varepsilon)$, or for $\tilde{\gamma}(\varepsilon) = \hat{\gamma}(\varepsilon)$, we have for the Gaussian errors ε:

$$Cov(\tilde{\gamma}; X) = Cov(\tilde{\gamma}; \varepsilon) = A\Sigma$$

and

$$Cov(\tilde{\gamma}; U) = Cov(\tilde{\gamma}; \varepsilon(n+1)) = Ar,$$

since the third moments of ε are equal to zero, from which we have

$$Cov(\tilde{\gamma}; X)\Sigma^{-1}r - Cov(\tilde{\gamma}; U)) = 0.$$

Thus for a Gaussian time series $X(.)$ we have for $\tilde{\gamma}(\varepsilon) = \hat{\gamma}_\Sigma(\varepsilon)$, or for $\tilde{\gamma}(\varepsilon) = \hat{\gamma}(\varepsilon)$ and for $U = X(n+d)$ and $\tilde{U} = m_{\tilde{\gamma}}(n+d) + r'\Sigma^{-1}(X - m_{\tilde{\gamma}})$, the approximation

$$
\begin{aligned}
E_{\gamma_0}[\tilde{U} - U]^2 &\approx D_{\gamma_0}[U] - r'\Sigma^{-1}r + \left\| f_{\gamma_0} - F'_{\gamma_0}\Sigma^{-1}r \right\|^2_{Cov_{\gamma_0}(\tilde{\gamma})} \\
&\quad + \left\| f_{\gamma_0} - F'_{\gamma_0}\Sigma^{-1}r \right\|^2_{b_0 b'_0} \\
&= D_{\gamma_0}[U] - r'\Sigma^{-1}r + \left\| f_{\gamma_0} - F'_{\gamma_0}\Sigma^{-1}r \right\|^2_{Cov_{\gamma_0}(\tilde{\gamma}) + b_0 b'_0},
\end{aligned}
$$

where the expressions for the biases b_0 and $Cov_{\gamma_0}(\tilde{\gamma})$ for $\tilde{\gamma} = \hat{\gamma}_\Sigma$ and for $\tilde{\gamma} = \hat{\gamma}$ are derived above. It should be expected that

$$Cov_{\gamma_0}(\hat{\gamma}_\Sigma) + b_0(\hat{\gamma}_\Sigma)b'_0(\hat{\gamma}_\Sigma) \leq Cov_{\gamma_0}(\hat{\gamma}) + b_0(\hat{\gamma})b'_0(\hat{\gamma})$$

and thus

$$E_{\gamma_0}[U^* - U]^2 \leq E_{\gamma_0}[\hat{U} - U]^2,$$

where

$$\hat{U} = m_{\hat{\gamma}}(n+d) + r'\Sigma^{-1}(X - m_{\hat{\gamma}}).$$

All the methods derived in this chapter are based on the assumption that the covariance matrix Σ and the vector r of covariances are known. This condition in usually not fulfilled in practical applications of time series and thus it is necessary to use also some model for the covariance function $R(.,.)$ of the observed time series $X(.)$. Using the model of covariance stationarity we can estimate the unknown covariance function $R(.)$ using the DOOLSEs $\hat{R}(.)$. We know that only estimates $\hat{R}(t)$ for small values of t have small MSE and thus only these values can be used as components of the estimated covariance matrix $\hat{\Sigma}$ and the vector $\hat{r}$. Thus a reasonable *empirical predictor* of $U = X(n+d)$ for this model of a covariance function is

$$\hat{U} = m_{\hat{\gamma}}(n+d) + \hat{r}'\hat{\Sigma}^{-1}(X - m_{\hat{\gamma}}),$$

where $\hat{\gamma}$ is the OLSE of γ, $\hat{\Sigma}$ is the estimated $p \times p$ covariance matrix of the last p components of X, and $\hat{r}$ and $X - m_{\hat{\gamma}}$ are the $p \times 1$ random vectors of the last p components of $\hat{r}$ and $X - m_{\hat{\gamma}}$, respectively.

In the case where we use some parametric model for $R(.,.)$, which means, if we assume that $R(.,.) \in \Xi = \{R_\nu(.,.); \nu \in \Upsilon\}$, that we can use the estimates $\tilde{\gamma}, \tilde{\nu}$ and define *the empirical predictor* $\tilde{U}$ by

$$\tilde{U} = m_{\tilde{\gamma}}(n + d) + \tilde{r}' \tilde{\Sigma}^{-1}(X - m_{\tilde{\gamma}}),$$

where $\tilde{r} = r_{\tilde{\nu}}$ and $\tilde{\Sigma} = \Sigma_{\tilde{\nu}}$.

All the methods derived in this chapter are based on the assumption that the covariance matrix Σ and the vector r of covariances are known. This condition in usually not fulfilled in the practical applications of time series. In the following chapter we derive predictors which will be based on the estimated covariance characteristics of time series $X(.)$.

5

Empirical Predictors

5.1 Introduction

In the preceding chapter we considered different predictors in models at which the mean value of $X(.)$ was unknown and modeled by some regression models. The main assumption under which these predictors were derived was that the covariance characteristics of the models were known. This is a rather strong assumption which is fulfilled only rarely in practical problems when we have usually only time series data x consisting of a finite realization of an observation X of $X(.)$ and nothing more. Thus it is necessary to model not only the mean value $m(.)$ of $X(.)$, but also the covariance function $R(.,.)$ of $X(.)$. The main models for covariance functions were considered in the preceding chapter and also the methods, giving the estimators, such as the DOOLSE, the DOWELSE, and the MLE, of parameters of covariance functions were derived.

Using the model of covariance stationarity we can estimate the unknown covariance function $R(.)$ using the DOOLSEs $\hat{R}(.)$. We know that only estimates $\hat{R}(t)$ for small values of t have small MSE, and thus only these values can be used as components of the estimated covariance matrix $\hat{\Sigma}$ and the vector $\hat{r}$. Thus the reasonable *empirical predictor* $\tilde{U}$ of $U = X(n + d)$ for this model of covariance function is

$$\tilde{U} = \hat{m}(n + d) + \hat{r}'\hat{\Sigma}^{-1}(X - \hat{m}),$$

where $\hat{m}(.)$ is the estimator of $m(.)$ based on the OLSEs $\hat{\beta}$ or $\hat{\gamma}$, $\hat{\Sigma}$ is the estimated $p \times p$ covariance matrix of the last p components of X and $\hat{r}$ and $X - \hat{m}$ are the $p \times 1$ random vectors of the last p components of $\hat{r}$ and $X - m_{\hat{\gamma}}$, respectively.

In the case when we use some parametric model for $R(.,.)$ that means, if we assume that $R(.,.) \in \Xi = \{R_\nu(.,.); \nu \in \Upsilon\}$, we can use the estimates $\tilde{\beta}$ or $\tilde{\gamma}$ and $\tilde{\nu}$ and define *the empirical predictor* $\tilde{U}$ by

$$\tilde{U} = \tilde{m}(n + d) + \tilde{r}'\tilde{\Sigma}^{-1}(X - \tilde{m}),$$

where $\tilde{m} = m_{\tilde{\beta}}$ or $\tilde{m} = m_{\tilde{\gamma}}$, $\tilde{r} = r_{\tilde{\nu}}$, and $\tilde{\Sigma} = \Sigma_{\tilde{\nu}}$.

The MSEs of these empirical predictors are not known, but it can be expected that they are greater than the MSEs of predictors with known covariance function. Also it can be expected that the MSE of the empirical predictor with the MLEs $\hat{\gamma}$ and $\tilde{\nu}$ will probably be smaller than the MSE of the empirical predictor with parameters $\hat{\gamma}$ and $\hat{\nu}$ estimated by the double least squares method. The derivation of an expression for the MSEs of these empirical predictors is an open problem. Some results on empirical predictors can be found in Eaton (1985), Harville (1985) and Christensen (1991).

5.2   Properties of Empirical Predictors

We shall study the properties of empirical predictors in linear LPRMs in the case when covariance functions of the observed time series depend on a covariance parameter.

Let $X = (X(1), ..., X(n))'$ be a finite observation of $X(.)$ and let $Z = (X', U)'$ be given by the prediction regression model

$$Z = \begin{pmatrix} X \\ U \end{pmatrix} = \begin{pmatrix} F \\ f' \end{pmatrix} \beta + \varepsilon; \beta \in E^k; E[\varepsilon] = 0,$$

where we assume that the covariance functions $R_\nu(.,.)$ of $X(.)$, and thus also the covariance vectors r_ν and covariance matrices $Cov_\nu(Z); \nu \in \Upsilon$, depend on the covariance parameter ν. Then we can write

$$Cov_\nu(Z) = \begin{pmatrix} \Sigma_\nu & r_\nu \\ r_\nu' & D_\nu[U] \end{pmatrix}; \nu \in \Upsilon.$$

Let $\tilde{\theta} = (\tilde{\beta}', \tilde{\nu}')'$ be some estimator of the unknown parameter $\theta = (\beta, \nu)'$ of the prediction regression model, let $U = X(n + d)$, and let

$$\tilde{U} = f'\tilde{\beta} + \tilde{r}'\tilde{\Sigma}^{-1}(X - F\tilde{\beta}),$$

where $\tilde{r} = r_{\tilde{\nu}}$ and $\tilde{\Sigma} = \Sigma_{\tilde{\nu}}$, be the empirical predictor of U. We shall assume further that

$$\tilde{\beta}(X) = (F'\tilde{\Sigma}^{-1}F)^{-1}F'\tilde{\Sigma}^{-1}X$$

is the empirical version of the BLUP β_Σ^* with the unknown covariance matrix Σ replaced by their estimator $\tilde{\Sigma} = \Sigma_{\tilde{\nu}}$.

We shall now show that under some conditions on the estimators $\tilde{r}$ and $\tilde{\Sigma}$ the empirical predictor $\tilde{U}$ is an unbiased predictor of U. This result is based on the following notion.

Let $\tilde{A}(X)$ be a random $n \times q$ matrix depending on X. Then $\tilde{A}(X)$ will be called a *residual-type statistics (RTSs)*, if

$$\tilde{A}(X) = \tilde{A}(X + F\beta) \text{ for all } \beta \in E^k$$

and

$$\tilde{A}(X) = \tilde{A}(-X).$$

Remarks. 1. For every RTS $\tilde{A}$ we can write $\tilde{A}(X) = \tilde{A}(\varepsilon) = \tilde{A}(-\varepsilon)$.

2. All estimators $\tilde{r} = r_{\tilde{\nu}}$ and $\tilde{\Sigma} = \Sigma_{\tilde{\nu}}$, based on the DOOLSE, the WELSE, or the MLE $\tilde{\nu}$ are RTSs, since they are functions of invariant quadratic forms for which the conditions appearing in the definition of the RTSs are fulfilled. As a consequence we also get that the statistics $(F'\tilde{\Sigma}^{-1}F)^{-1}F'\tilde{\Sigma}^{-1}, \tilde{r}'\tilde{\Sigma}^{-1}, M_{\tilde{\Sigma}} = I - P_{\tilde{\Sigma}} = I - F(F'\tilde{\Sigma}^{-1}F)^{-1}F'\tilde{\Sigma}^{-1}$, and $\tilde{r}'\tilde{\Sigma}^{-1}M_{\tilde{\Sigma}}$ are RTSs if $\tilde{\nu}$ is a function of invariant quadratic forms, which we shall assume in all of this section.

The following lemma is a key for showing the unbiasedness of $\tilde{U}$.

Lemma 5.2.1. *Let $\tilde{A}(X)$ be an RTS. Then*

$$E[\tilde{A}(X)X] = E[\tilde{A}(X)F\beta].$$

Proof. We have

$$E[\tilde{A}(X)X] = E[\tilde{A}(X)F\beta] + E[\tilde{A}(X)\varepsilon] = E[\tilde{A}(X)F\beta],$$

since

$$E[\tilde{A}(X)\varepsilon] = E[\tilde{A}(\varepsilon)\varepsilon] = -E[\tilde{A}(-\varepsilon)\varepsilon] = -E[\tilde{A}(\varepsilon)\varepsilon]$$

and thus $E[\tilde{A}(X)\varepsilon] = 0$.

As a direct consequence of this lemma we get that $\tilde{\beta}(X)$ *is an unbiased estimator of β if $\tilde{\Sigma}$ is an RTS.* Now we can prove the following theorem.

Theorem 5.2.1. *Let*

$$\tilde{U}(X) = f'\tilde{\beta} + \tilde{r}'\tilde{\Sigma}^{-1}(X - F\tilde{\beta}),$$

where $\tilde{\beta}(X) = (F'\tilde{\Sigma}^{-1}F)^{-1}F'\tilde{\Sigma}^{-1}X$ and $\tilde{r}$ and $\tilde{\Sigma}$ are RTSs. Then $\tilde{U}$ is an unbiased predictor of U.

Proof. Since $\tilde{\beta}$ is an unbiased estimator of β, it is enough to prove that

$$E[\tilde{r}'\tilde{\Sigma}^{-1}(X - F\tilde{\beta})] = 0.$$

We can write

$$\tilde{r}'\tilde{\Sigma}^{-1}(X - F\tilde{\beta}) = \tilde{r}'\tilde{\Sigma}^{-1}(I - P_{\tilde{\Sigma}})X = \tilde{r}'\tilde{\Sigma}^{-1}M_{\tilde{\Sigma}}X,$$

where $\tilde{r}'\tilde{\Sigma}^{-1}M_{\tilde{\Sigma}}$ is an RTS and thus, using Lemma 5.2.1,

$$E[\tilde{r}'\tilde{\Sigma}^{-1}(X - F\tilde{\beta})] = E[\tilde{r}'\tilde{\Sigma}^{-1}M_{\tilde{\Sigma}}F\beta] = 0,$$

since $P_{\tilde{\Sigma}}$ is a projector on $L(F)$ and thus $P_{\tilde{\Sigma}}F = F$ and

$$M_{\tilde{\Sigma}}F = (I - P_{\tilde{\Sigma}})F = 0.$$

The theorem is proved.

Let Σ be a given covariance matrix, let $M_\Sigma = I - P_\Sigma$, where $P_\Sigma = F(F'\Sigma^{-1}F)^{-1}F'\Sigma^{-1}$, and let $U_\Sigma^*(X) = f'\beta_\Sigma^*(X) + r'\Sigma^{-1}(X - F\beta_\Sigma^*(X))$ be the BLUP for U by the given Σ. We now derive an expression for $U_\Sigma^* - \tilde{U}$. We can write

$$\tilde{U}(X) = \tilde{a}'X,$$

where

$$\begin{aligned}
\tilde{a}' &= f'(F'\tilde{\Sigma}^{-1}F)^{-1}F'\tilde{\Sigma}^{-1} + \tilde{r}'\tilde{\Sigma}^{-1}M_{\tilde{\Sigma}} \\
&= g'P_{\tilde{\Sigma}} + \tilde{r}'\tilde{\Sigma}^{-1}M_{\tilde{\Sigma}},
\end{aligned}$$

since we can write $f = F'g$ for some $g \in E^n$, if $r(F) = k$. Next we have

$$\tilde{U}(X) = \tilde{a}'P_\Sigma X + \tilde{a}'(I - P_\Sigma)X$$

and

$$\tilde{a}'P_\Sigma X = g'P_{\tilde{\Sigma}}P_\Sigma X + \tilde{r}'\tilde{\Sigma}^{-1}M_{\tilde{\Sigma}}P_\Sigma X = g'P_\Sigma X = f'\beta_\Sigma^*(X),$$

since $P_{\tilde{\Sigma}}$ and P_Σ are projectors on $L(F)$ and thus

$$P_{\tilde{\Sigma}}P_\Sigma = P_\Sigma \quad \text{and} \quad M_{\tilde{\Sigma}}P_\Sigma = (I - P_{\tilde{\Sigma}})P_\Sigma = 0.$$

Thus we have

$$\begin{aligned}
\tilde{U}(X) &= f'\beta_\Sigma^*(X) + \tilde{a}'M_\Sigma X \\
&= f'\beta_\Sigma^*(X) + r'\Sigma^{-1}M_\Sigma X + (\tilde{a}' - r'\Sigma^{-1})M_\Sigma X,
\end{aligned}$$

from which

$$U_\Sigma^* - \tilde{U} = (r'\Sigma^{-1} - \tilde{a}')M_\Sigma X.$$

Here

$$M_\Sigma X = X - F\beta_\Sigma^*(X) = \hat{\varepsilon}_\Sigma(X)$$

are *the weighted least squares residuals* and we see that $U_\Sigma^* - \tilde{U}$ is a function of the weighted residuals $\hat{\varepsilon}_\Sigma(X)$, if $\tilde{a}(X)$ is also a function of the weighted residuals, which is true if $\tilde{\Sigma}(X)$ and $\tilde{r}(X)$ can also be regarded as functions of the weighted residuals $\hat{\varepsilon}_\Sigma(X)$.

Let P_F be any deterministic projector on $L(F)$ and let $M_F = I - P_F$. Then, since $P_F P_\Sigma = P_\Sigma$ and $M_F P_\Sigma = 0$, we can write

$$M_F M_\Sigma = M_F(I - P_\Sigma) = M_F - M_F P_\Sigma = M_F$$

and we see that any residuals $M_F X$ can be written as

$$M_F X = M_F M_\Sigma X = M_F \hat{\varepsilon}_\Sigma(X).$$

Thus they are a function of the weighted residuals $\hat{\varepsilon}_\Sigma(X)$. Especially, for $M_F = M = I - F(F'F)^{-1}F'$, we get that *the ordinary least squares residuals*

$$\hat{\varepsilon}(X) = MX$$

can be written in the form

$$\hat{\varepsilon}(X) = M\hat{\varepsilon}_\Sigma(X)$$

and we see that they can be considered as a function of the weighted residuals.

Thus, in the case when $\tilde{r} = r_{\tilde{\nu}}$ and $\tilde{\Sigma} = \Sigma_{\tilde{\nu}}$ are based on the DOOLSE $\tilde{\nu}$, which is a linear function of the ordinary residuals, we can state that $\tilde{a}$ is a function of weighted residuals.

In the case when $\tilde{r} = r_{\tilde{\nu}}$ and $\tilde{\Sigma} = \Sigma_{\tilde{\nu}}$ are based on the MLE $\tilde{\nu}$, which is computed iteratively with the initial value, the DOOLSE $\nu^{(0)}$, which is a linear function of the ordinary residuals, we can state that all iterations $\nu^{(i)}; i = 0, 1, \ldots$ as functions of $\nu^{(i-1)}$ are functions of the weighted residuals and thus $\tilde{a}$ is also a function of the weighted residuals.

If we set $M_F = M_{\tilde{\Sigma}}$, then we have

$$M_{\tilde{\Sigma}} = M_{\tilde{\Sigma}} M_\Sigma$$

and *the empirical residuals* $M_{\tilde{\Sigma}} X = \tilde{\varepsilon}(X)$ will be given by a function of the weighted residuals, if $M_{\tilde{\Sigma}}$ is a function of the weighted residuals.

The following theorem gives an expression for the MSE of an empirical predictor $\tilde{U}$.

Theorem 5.2.2. *Let Σ be a given covariance matrix, let $U_\Sigma^*(X)$ be the BLUP of U, and let*

$$\tilde{U} = f'\tilde{\beta} + \tilde{r}'\tilde{\Sigma}^{-1}(X - F\tilde{\beta}),$$

where $\tilde{r}$ and $\tilde{\Sigma}$ are RTSs which are functions of the weighted residuals $\hat{\varepsilon}_\Sigma(X) = M_\Sigma X$. Let $E_\Sigma[U - U_\Sigma^* \mid \hat{\varepsilon}_\Sigma(X)] = 0$. Then

$$E_\Sigma[\tilde{U} - U]^2 = E_\Sigma[U_\Sigma^* - U]^2 + E_\Sigma[U_\Sigma^* - \tilde{U}]^2.$$

Proof. We can write

$$E_\Sigma[U - \tilde{U}]^2 = E_\Sigma[U - U_\Sigma^*]^2 + E_\Sigma[U_\Sigma^* - \tilde{U}]^2 + 2Cov_\Sigma(U - U_\Sigma^*; U_\Sigma^* - \tilde{U})$$

and the theorem will be proved if we prove that $Cov_\Sigma(U - U^*_\Sigma; U^*_\Sigma - \tilde{U}) = 0$. But

$$
\begin{aligned}
Cov_\Sigma(U - U^*_\Sigma; U^*_\Sigma - \tilde{U}) &= E_\Sigma[(U - U^*_\Sigma)(U^*_\Sigma - \tilde{U})] \\
&= E_\Sigma[E_\Sigma[(U - U^*_\Sigma)(U^*_\Sigma - \tilde{U}) \mid \hat{\varepsilon}_\Sigma(X)]] \\
&= E_\Sigma[(U^*_\Sigma - \tilde{U})E_\Sigma[U - U^*_\Sigma \mid \hat{\varepsilon}_\Sigma(X)]] = 0
\end{aligned}
$$

and the theorem is proved.

As a consequence of this theorem we have the inequality

$$
E_\Sigma[\tilde{U} - U]^2 \geq E_\Sigma[U^*_\Sigma - U]^2,
$$

which holds under the assumptions of the preceding theorem and which says that the MSE of the empirical predictor $\tilde{U}$ is greater than the MSE of the BLUP U^*_Σ.

Now we shall show that the condition $E_\Sigma[U - U^*_\Sigma \mid \hat{\varepsilon}_\Sigma(X)] = 0$ holds if $Z = (X', U)'$ has a multivariate normal distribution with covariance matrix

$$
Cov(Z) = \begin{pmatrix} \Sigma & r \\ r' & D[U] \end{pmatrix}.
$$

We can write

$$
\begin{aligned}
Cov_\Sigma(U - U^*_\Sigma; \hat{\varepsilon}_\Sigma(X)) &= Cov_\Sigma(U; M_\Sigma X) - Cov_\Sigma(U^*_\Sigma; M_\Sigma X) \\
&= M_\Sigma Cov_\Sigma(U; X) - M_\Sigma Cov_\Sigma(U^*_\Sigma; X) \\
&= M_\Sigma r - M_\Sigma Cov_\Sigma(f'\beta^*_\Sigma + r'\Sigma^{-1}M_\Sigma X; X) \\
&= M_\Sigma r - M_\Sigma Cov_\Sigma(f'\beta^*_\Sigma; X) \\
&\quad - M_\Sigma Cov_\Sigma(r'\Sigma^{-1}M_\Sigma X; X).
\end{aligned}
$$

Next we have

$$
M_\Sigma Cov_\Sigma(f'\beta^*_\Sigma; X) = M_\Sigma Cov_\Sigma(g'P_\Sigma X; X) = M_\Sigma \Sigma P'_\Sigma g = 0,
$$

since

$$
\begin{aligned}
M_\Sigma \Sigma P'_\Sigma &= (I - P_\Sigma)\Sigma P'_\Sigma = \Sigma P'_\Sigma - P_\Sigma \Sigma P'_\Sigma \\
&= F(F'\Sigma^{-1}F)^{-1}F' - F(F'\Sigma^{-1}F)^{-1}F' = 0
\end{aligned}
$$

and

$$
M_\Sigma Cov_\Sigma(r'\Sigma^{-1}M_\Sigma X; X) = M_\Sigma \Sigma M'_\Sigma \Sigma^{-1}r = M_\Sigma r,
$$

since

$$M_\Sigma \Sigma M_\Sigma' = M_\Sigma \Sigma (I - P_\Sigma') = M_\Sigma \Sigma.$$

Thus we have proved that $U - U_\Sigma^*$, and the weighted residuals $\hat{\varepsilon}_\Sigma$, are uncorrelated and, since they are normally distributed, they are independent random variables. We have

$$E_\Sigma[U - U_\Sigma^* \mid \hat{\varepsilon}_\Sigma(X)] = E_\Sigma[U - U_\Sigma^*] = 0,$$

since the BLUP U_Σ^* is an unbiased predictor of U.

An approximation for $\tilde{U} - U_\Sigma^*$ was also derived in Kackar and Harville (1984). Since $\tilde{U} = \tilde{U}(X, \tilde{\nu})$ is a function of an estimator $\tilde{\nu}$ of ν, we can expand $\tilde{U}(X, \tilde{\nu})$ in a Taylor series in $\tilde{\nu}$ about the true value ν_0 and we obtain

$$\tilde{U} \approx U_\Sigma^* + \frac{\partial \tilde{U}(X, \nu)}{\partial \nu'} \mid_{\nu=\nu_0} (\tilde{\nu} - \nu_0),$$

from which

$$\tilde{U} - U_\Sigma^* \approx \frac{\partial \tilde{U}(X, \nu)}{\partial \nu'} \mid_{\nu=\nu_0} (\tilde{\nu} - \nu_0).$$

Thus the MSE of $\tilde{U} - U_\Sigma^*$ can be approximated by

$$E_{\nu_0}[\tilde{U} - U_{\nu_0}^*]^2 \approx E_{\nu_0} \left[\frac{\partial \tilde{U}(X, \nu)}{\partial \nu'} \mid_{\nu=\nu_0} (\tilde{\nu} - \nu_0) \right]^2.$$

The exact expression for the value on the right-hand side of the approximation is not known. Since $\tilde{\nu}$ is a function of quadratic forms in X and $\partial \tilde{U}(X, \nu)/\partial \nu' \mid_{\nu=\nu_0}$ is a linear function of X this depends on the sixth moments function of X. In the case when $Z = (X', U)'$ has a multivariate normal distribution, the sixth moments can be expressed as functions of the covariance function $R_{\nu_0}(.,.)$.

Kackar and Harville (1984) use the approximation

$$E_{\nu_0} \left[\frac{\partial \tilde{U}(X, \nu)}{\partial \nu'} \mid_{\nu=\nu_0} (\tilde{\nu} - \nu_0) \right]^2 \approx tr \left(Cov_{\nu_0} \left(\frac{\partial \tilde{U}(X, \nu)}{\partial \nu'} \mid_{\nu=\nu_0} \right) Cov_{\nu_0}(\tilde{\nu}) \right)$$

and give an expression for $Cov_\nu(\partial \tilde{U}(X, \nu)/\partial \nu')$. They showed that the ijth element c_{ij} of this covariance matrix is $c_{ij} = c_i' \Sigma_\nu^{-1} M_\nu c_j; i, j = 1, 2, ..., l$, where

$$c_i' = \frac{\partial r_\nu'}{\partial \nu_i} - (r_\nu' \Sigma_\nu^{-1} M_\nu + f'(F' \Sigma_\nu^{-1} F)^{-1} F' \Sigma_\nu^{-1}) \frac{\partial \Sigma_\nu}{\partial \nu_i}.$$

The asymptotic covariance matrix for the MLE $\tilde{\nu}$ was derived in a preceding section of this book.

We can consider the random variable

$$\tilde{E}[\tilde{U} - U]^2 = \tilde{D}[U] - \tilde{r}'\tilde{\Sigma}^{-1}\tilde{r} + (f - F'\tilde{\Sigma}^{-1}\tilde{r})'(F'\tilde{\Sigma}^{-1}F)^{-1}(f - F'\tilde{\Sigma}^{-1}\tilde{r})$$

as an estimator of the MSE of the empirical predictor $\tilde{U}$. We shall show now that this estimator underestimates the true value of the MSE of $\tilde{U}$. To show this let

$$\mathcal{U} = \{a \in E^n : a'F = f'\}.$$

Then any unbiased predictor $\hat{U}$ of U can be written as $\hat{U}(X) = a'X$. Let us use the notation

$$Cov((X', U)') = \begin{pmatrix} \Sigma & r \\ r' & D[U] \end{pmatrix} = V$$

and let Ξ denote a set of all p.d. V. Let us define, for every $a \in \mathcal{U}$, the function $g_a(.)$ defined on Ξ by

$$g_a(V) = E_V[a'X - U]^2; V \in \Xi.$$

Then it is clear that

$$g_a(V) = D_V[U] - 2a'r + a'\Sigma a$$

and, for $U = X(n + d)$,

$$g_a(\alpha V_1 + (1 - \alpha)V_2) = \alpha g_a(V_1) + (1 - \alpha)g_a(V_2).$$

Thus $g_a(.)$ is, for every given $a \in \mathcal{U}$, a concave function of V defined on Σ. We remark that some $a \in \mathcal{U}$ can be functions of V, but for a given $a_0 = a(V_0)$, $g_{a_0}(.)$ is a concave function of V.

Now let us consider a function $g(.)$ defined on Ξ by

$$g(V) = \min_{a \in \mathcal{U}} g_a(V); V \in \Xi.$$

It can be easily shown that $g(.)$, as a minimum of concave functions, is a concave function.

From the definition of the BLUP we get that

$$g(V) = D[U] - r'\Sigma^{-1}r + (f - F'\Sigma^{-1}r)'(F'\Sigma^{-1}F)^{-1}(f - F'\Sigma^{-1}r).$$

Let us assume now that

$$\tilde{V} = \begin{pmatrix} \tilde{\Sigma} & \tilde{r} \\ \tilde{r}' & \tilde{D}[U] \end{pmatrix}$$

is an unbiased estimator of V. Then $g(\tilde{V})$ is a random variable and, using the *Jensen inequality*,

$$g(E_V[\tilde{V}]) \geq E_V[g(\tilde{V})],$$

which holds for every concave function $g(.)$, we get

$$\begin{aligned} g(V) \quad \geq \quad & E_V[g(\tilde{V})] = E_V[\tilde{D}[U] - \tilde{r}'\tilde{\Sigma}^{-1}\tilde{r} \\ & + (f - F'\tilde{\Sigma}^{-1}\tilde{r})'(F'\tilde{\Sigma}^{-1}F)^{-1}(f - F'\tilde{\Sigma}^{-1}\tilde{r})]. \end{aligned}$$

Since

$$g(V) = E_V[U_V^* - U]^2,$$

where U_V^* is the BLUP of U by the given V, we get, under the assumptions of Theorem 5.2.2, the inequality

$$\begin{aligned} E_V[\tilde{U} - U]^2 \quad \geq \quad & g(V) \geq E_V[\tilde{D}[U] - \tilde{r}'\tilde{\Sigma}^{-1}\tilde{r} \\ & + (f - F'\tilde{\Sigma}^{-1}\tilde{r})'(F'\tilde{\Sigma}^{-1}F)^{-1}(f - F'\tilde{\Sigma}^{-1}\tilde{r})]. \end{aligned}$$

We note that the MLE $\hat{V}$ of V is asymptotically unbiased and for this estimator the assumptions of Theorem 5.2.2 are fulfilled and thus the last inequality is true asymptotically. This says that with a high probability

$$\tilde{D}[U] - \tilde{r}'\tilde{\Sigma}^{-1}\tilde{r} + (f - F'\tilde{\Sigma}^{-1}\tilde{r})'(F'\tilde{\Sigma}^{-1}F)^{-1}(f - F'\tilde{\Sigma}^{-1}\tilde{r}) \leq E_V[\tilde{U} - U]^2$$

and thus the value on the left-hand side of the last inequality underestimates the true MSE of $\tilde{U}$.

Harville and Jeske (1992) showed that for an unbiased estimator $\tilde{V}$ of V:

$$E_V[\tilde{U} - U_V^*]^2 = E_V[U_V^* - U]^2 - E_V[g(\tilde{V})] = g(V) - E_V[g(\tilde{V})].$$

Using a second-order Taylor series expansion of $g(.)$ we get the approximation

$$g(\tilde{V}) \approx g(V_{\nu_0}) + \frac{\partial g(V_\nu)}{\partial \nu'} \mid_{\nu=\nu_0} (\tilde{\nu} - \nu_0) + \frac{1}{2}(\tilde{\nu} - \nu_0)' \frac{\partial^2 g(V_\nu)}{\partial \nu \partial \nu'} \mid_{\nu=\nu_0} (\tilde{\nu} - \nu_0)$$

from which, for the unbiased estimator $\tilde{\nu}$ of ν, we get

$$E_{\nu_0}[g(\tilde{V})] \approx g(V_{\nu_0}) + \frac{1}{2}tr\left(\frac{\partial^2 g(V_\nu)}{\partial \nu \partial \nu'} \mid_{\nu=\nu_0} E_{\nu_0}[(\tilde{\nu} - \nu_0)(\tilde{\nu} - \nu_0)']\right).$$

Using the statement of Theorem 5.2.2 and the last approximation we can write, under the assumptions of Theorem 5.2.2, the approximation

$$E_V[\tilde{U} - U]^2 \approx E_V[U_V^* - U]^2 - \frac{1}{2}tr\left(\frac{\partial^2 g(V_\nu)}{\partial \nu \partial \nu'} Cov_\nu(\tilde{\nu})\right).$$

It is possible to show that if V_ν is linear in ν, then

$$\frac{\partial^2 g(V_\nu)}{\partial \nu \partial \nu'} = -2Cov_\nu \left(\frac{\partial \tilde{U}(X, \nu)}{\partial \nu'} \right)$$

and we get the approximation

$$E_V [\tilde{U} - U]^2 \approx E_V [U_V^* - U]^2 + tr \left(Cov_\nu \left(\frac{\partial \tilde{U}(X, \nu)}{\partial \nu'} \right) Cov_\nu(\tilde{\nu}) \right),$$

which is identical with that derived by Kackar and Harville (1984).

Example 5.2.1. Let us consider a Gaussian stationary time series $X(.)$ given by the model

$$X(t) = \beta + \varepsilon(t); t = 1, 2, ...,$$

with covariance function $R_\nu(t) = \left[\sigma^2 / \left(1 - \rho^2 \right) \right] \rho^t; t = 0, 1, ..., \nu = (\rho, \sigma^2)$. Then, see Example 4.2.2,

$$E_\nu [X^*(n+1) - X(n+1)]^2 = g(V_\nu) = \sigma^2 \left(1 + \frac{(1-\rho)}{2 + (n-2)(1-\rho)} \right)$$

and we get, after some computation,

$$\frac{\partial^2 g(V_\nu)}{\partial \rho^2} = \frac{-4\sigma^2(n-2)}{[2 + (n-2)(1-\rho)]^3},$$

$$\frac{\partial^2 g(V_\nu)}{\partial \sigma^2 \partial \rho} = \frac{-2\sigma^2}{[2 + (n-2)(1-\rho)]^2},$$

$$\frac{\partial^2 g(V_\nu)}{(\partial \sigma^2)^2} = 0.$$

If $\tilde{\nu} = (\tilde{\rho}, \tilde{\sigma}^2)$ is the MLE of ν, then, see Example 3.4.4, the asymptotic covariance matrix of $\tilde{\nu}$ is

$$Cov_\nu(\tilde{\nu}) \approx \frac{1}{n} \left(\begin{array}{cc} 1 - \rho^2 & 0 \\ 0 & 2\sigma^4 \end{array} \right).$$

Thus, for large n we can write the approximation

$$E_\nu [\tilde{X}(n+1) - X(n+1)]^2 \approx \sigma^2 \left(1 + \frac{(1-\rho)}{2 + (n-2)(1-\rho)} \right)$$
$$+ \frac{2\sigma^2(n-2)(1-\rho^2)}{n[2 + (n-2)(1-\rho)]^3}$$

and

$$\lim_{n\to\infty} E_\nu[\tilde{X}(n+1) - X(n+1)]^2 \approx \sigma^2.$$

Example 5.2.2. Let us consider the time series

$$X(t) = \beta + \varepsilon(t); t = 1, 2, ...; \beta \in E^1,$$

where $\varepsilon(.)$ is a time series with covariance functions

$$R_\nu(t) = \sigma^2 e^{-\alpha t}; t = 0, 1, 2, ...,$$

with $\nu = (\sigma^2, \alpha)' \in (0, \infty) \times (0, \infty)$. Then $F = (1, 1, ..., 1)'$, $f_{(n)} = f = 1$, and, see Example 4.2.4,

$$
\begin{aligned}
Cov(\beta_\nu^*) &= \frac{\sigma^2(1 - e^{-2\alpha})}{n + 2(n-1)e^{-\alpha} + (n-2)e^{-2\alpha}} \\
&= \frac{\sigma^2(1 - e^{-2\alpha})}{2(1 - e^{-\alpha}) + (n-2)(1 - e^{-\alpha})^2},
\end{aligned}
$$

$$
\begin{aligned}
MSE_\nu[X^*(n+d)] &= \sigma^2(1 - e^{-2\alpha d}) \\
&\quad + (1 - e^{-\alpha d})^2 \frac{\sigma^2(1 - e^{-2\alpha})}{2(1 - e^{-\alpha}) + (n-2)(1 - e^{-\alpha})^2}.
\end{aligned}
$$

Thus

$$
\begin{aligned}
g(V_\nu) &= MSE_\nu[X^*(n+1)] \\
&= \sigma^2(1 - e^{-2\alpha})\left(1 + \frac{(1 - e^{-\alpha})^2}{2(1 - e^{-\alpha}) + (n-2)(1 - e^{-\alpha})^2}\right) \\
&= \sigma^2(1 - e^{-2\alpha})\left(1 + \frac{1 - e^{-\alpha}}{2 + (n-2)(1 - e^{-\alpha})}\right).
\end{aligned}
$$

It is not difficult to compute the matrix $\partial^2 g(V_\nu)/\partial\nu\partial\nu'$, but for simplicity we shall compute only terms which do not vanish by large n. We can write

$$
\frac{\partial^2 g(V_\nu)}{(\partial\sigma^2)^2} = 0,
$$

$$
\frac{\partial^2 g(V_\nu)}{\partial\sigma^2 \partial\alpha} \approx 2e^{-2\alpha},
$$

$$
\frac{\partial^2 g(V_\nu)}{\partial\alpha^2} \approx -4\sigma^2 e^{-2\alpha}.
$$

If $\tilde{\nu} = (\tilde{\sigma}^2, \tilde{\alpha})$ is the MLE of ν, then, see Example 3.4.5, the asymptotic covariance matrix of $\tilde{\nu}$ is

$$Cov_\nu(\tilde{\nu}) = 2G_\nu^{-1} \approx \frac{1}{n} \begin{pmatrix} 2\sigma^4 \left(1 + e^{-2\alpha}\right) / \left(1 - e^{-2\alpha}\right) & -2\sigma^2 \\ -2\sigma^2 & \left(1 - e^{-2\alpha}\right) / e^{-2\alpha} \end{pmatrix}$$

and, for large n,

$$E_\nu[\tilde{X}(n+1) - X(n+1)]^2 \approx \sigma^2(1 - e^{-2\alpha}) \left(1 + \frac{1 - e^{-\alpha}}{2 + (n-2)(1 - e^{-\alpha})}\right)$$
$$- \frac{1}{2n} \left(-8\sigma^2 e^{-2\alpha} - 4\sigma^2(1 - e^{-2\alpha})\right)$$
$$\approx \sigma^2(1 - e^{-2\alpha}) \left(1 + \frac{1 - e^{-\alpha}}{2 + (n-2)(1 - e^{-\alpha})}\right)$$
$$+ \frac{2\sigma^2}{n}(1 + e^{-2\alpha}).$$

We have

$$\lim_{n \to \infty} E_\nu[\tilde{X}(n+1) - X(n+1)]^2 \approx \sigma^2(1 - e^{-2\alpha}).$$

It is an open problem to study properties of empirical predictors in multivariate and NRMs.

5.3   Numerical Examples

We shall now give a few examples to show the applications of kriging theory to real data. In most of these examples we use the following model

$$X(t) = \beta_1 + \beta_2 t + \sum_{i=1}^{k} (\beta_i^1 \cos \lambda_i t + \beta_i^2 \sin \lambda_i t) + \varepsilon(t); t = 1, 2, ..., n,$$

on real data. The mean value $m_\gamma(.)$ is assumed to be a linear trend plus a quasiperiodic component with some (possibly unknown) frequencies. The random error time series $\varepsilon(.)$ is assumed to be some stationary AR(p) time series.

Example 5.3.1. In the following table there are quarterly observed consumptions of some nonalcoholic drinks during 4 years:

t	$x(t)$	t	$x(t)$	t	$x(t)$	t	$x(t)$
1	21.4	5	23.1	9	23.5	13	25.2
2	33.2	6	33.5	10	34.2	14	36.0
3	40.8	7	41.8	11	42.3	15	43.4
4	31.2	8	31.8	12	30.4	16	35.5

For these data the length of the expected period is $T_p = 4$ and to this period corresponds the frequency $\lambda_p = 2\pi/4 = \pi/2$. This is also confirmed by the periodogram computed from the data with the eliminated linear trend given in Figure 5.3.1.

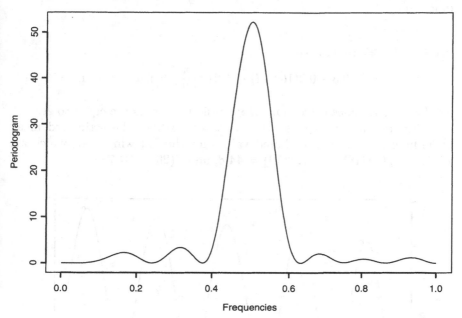

Figure 5.3.1.

According to this, we use for the data the LRM

$$X(t) = \beta_1 + \beta_2 t + \beta_1^1 \cos \frac{\pi}{2} t + \beta_2^2 \sin \frac{\pi}{2} t + \varepsilon(t); t = 1, 2, ...,$$

with unknown regression parameter $\beta = (\beta_1, \beta_2, \beta_1^1, \beta_2^2)'$. We estimate this parameter by the OLSE $\hat{\beta}$ and we get $\hat{\beta} = (30.8, 0.2, -1.2, -9.1)$. The estimated mean value function is

$$m_{\hat{\beta}}(t) = 30.8 + 0.2t - 1.2 \cos \frac{\pi}{2} t - 9.1 \sin \frac{\pi}{2} t; t = 1, 2, ...$$

We can estimate the values of the covariance function $R(.)$ by the DOOLSEs $\hat{R}(t); t = 0, 1, ..., 15$. Let

$$\hat{\varepsilon}(t) = x(t) - m_{\hat{\beta}}(t); t = 1, 2, ..., 16,$$

then

$$\hat{R}(t) = \frac{1}{16 - t} \sum_{s=1}^{16-t} \hat{\varepsilon}(s + t)\hat{\varepsilon}(s); t = 0, 1, ..., 15.$$

We get that $\hat{R}(0) = 0.6$ and other values of $\hat{R}(t); t = 1, ..., 15$, are practically equal to zero. We can see from this result that the time series $\varepsilon(.)$ can be regarded as a white noise. Using this assumption we can write the expressions for the predicted values as values of the prolonged mean value function

$$
\begin{aligned}
\hat{x}(16 + d) &= m_{\hat{\beta}}(16 + d) \\
&= 30.8 + 0.2(16 + d) - 1.2\cos\frac{\pi}{2}(16 + d) - 9.1\sin\frac{\pi}{2}(16 + d)
\end{aligned}
$$

for $d = 1, 2, ...$, since corrections from residuals terms are equal to zero.

In Figure 5.3.2. we can see the observed values, the estimated mean value function, and the predicted values for the following year, which are $\hat{x}(17) = 26.0, \hat{x}(18) = 36.7, \hat{x}(19) = 44.8$, and $\hat{x}(20) = 34.7$.

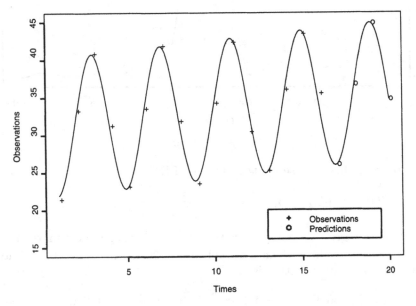

Figure 5.3.2.

Example 5.3.2. The following data are monthly observed incomes, in millions of crowns, for some company during 3 years.

t	$x(t)$	t	$x(t)$	t	$x(t)$
1	240	13	502	25	733
2	297	14	582	26	841
3	287	15	535	27	827
4	274	16	521	28	738
5	226	17	387	29	546
6	222	18	388	30	541
7	312	19	532	31	718
8	395	20	655	32	943
9	443	21	759	33	1048
10	450	22	680	34	983
11	341	23	515	35	722
12	480	24	756	36	946

We use the NRM

$$
\begin{aligned}
X(t) &= \beta_1 + \beta_2 t + \sum_{i=1}^{k} \left(\beta_i^1 \cos \lambda_i t + \beta_i^2 \sin \lambda_i t \right) + \varepsilon(t) \\
&= m_\gamma(t) + \varepsilon(t); t = 1, 2, \ldots,
\end{aligned}
$$

with an unknown k and with an unknown regression parameter γ. To find the OLSE $\hat{\gamma}$ we have to use some iterative procedure and we want to give some starting value $\hat{\gamma}^{(0)}$ of the iterations. This was done as follows. First the OLSE $\tilde{\beta} = (\tilde{\beta}_1, \tilde{\beta}_2)'$ of $\beta = (\beta_1, \beta_2)'$ was computed from the observed vector $x = (x(1), x(2), \ldots, x(36))'$. Let us denote by $\tilde{\varepsilon}$ the vector of residuals with components

$$
\tilde{\varepsilon}(t) = x(t) - \tilde{\beta}_1 - \tilde{\beta}_2 t; t = 1, 2, \ldots, 36.
$$

The periodogram of the residuals $\tilde{\varepsilon}$ was computed and it can be seen that $k = 2$ and that the frequencies $\lambda_1 = 2\pi/6$ and $\lambda_2 = 2\pi/12$ are significant. We can see this periodogram in Figure 5.3.3. These frequencies correspond to a year and to a half of the year periodicity, since $T_1 = 2\pi/\lambda_1 = 6$ and $T_2 = 2\pi/\lambda_2 = 12$.

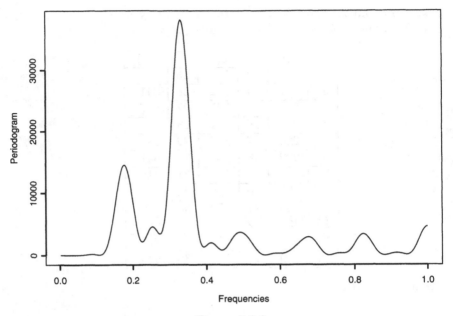

Figure 5.3.3.

Then the ordinary least squares method was used to find the OLSE $\hat{\gamma}$. We get

$$\hat{\beta}_1 = 200.4, \hat{\beta}_2 = 19.7,$$

$$\hat{\lambda}_1 = 0.33\pi, \hat{\beta}_1^1 = -95.5, \hat{\beta}_2^1 = 52.3,$$

$$\hat{\lambda}_2 = 0.16\pi, \hat{\beta}_2^1 = 61.6, \hat{\beta}_2^2 = -14.5.$$

The estimated mean value function is

$$
\begin{aligned}
m_{\hat{\gamma}}(t) \quad = \quad & 200.4 + 19.7t \\
& -95.5 \cos \frac{\pi}{3}t + 52.3 \sin \frac{\pi}{3}t \\
& +61.6 \cos \frac{\pi}{6}t - 14.5 \sin \frac{\pi}{6}t.
\end{aligned}
$$

Using this mean value function, the vector of the ordinary least squares residuals $\hat{\varepsilon}$ with components

$$\hat{\varepsilon}(t) = x(t) - m_{\hat{\gamma}}(t); t = 1, 2, ..., 36,$$

was used to find the DOOLSE $\hat{R}(.)$ of the covariance function $R(.)$ using the expression

$$\hat{R}(t) = \frac{1}{36-t} \sum_{s=1}^{36-t} \hat{\varepsilon}(s+t)\hat{\varepsilon}(s); t = 0, 1, ..., 35.$$

Some values of this estimator are used in the following table:

t	0	1	2	3	4	5	6	7	8	9
$\hat{R}(t)$	3242	-272	-511	-888	-475	-92	280	80	637	-88

In this case we do not assume that the error time series $\varepsilon(.)$ is a white noise, since the $\hat{R}(t)$ are not near to zero for $t = 1, 2, ..., 8$. For predicting the following values we shall assume that $\varepsilon(.)$ is an AR(8) time series. The predicted values for the next year were computed according to the expression

$$\hat{x}(36+d) = m_{\hat{\gamma}}(36+d) + \hat{r}_{d,(8)}\hat{\Sigma}_{(8)}^{-1}(x_{(8)} - m_{\hat{\gamma},(8)}); d = 1, 2, ..., 12,$$

where $\hat{r}_{d,(8)}$ is the 8×1 vector with components $\hat{R}(t+d-1); t = 1, 2, ..., 8, d = 1, 2, ..., 12, \hat{\Sigma}_{(8)}$ is the 8×8 matrix with components $\hat{\Sigma}_{(8),ij} = \hat{R}(i-j); i, j = 1, 2, ..., 8$, and the 8×1 vector $x_{(8)} - m_{\hat{\gamma},(8)}$ consists of the last eight components of the residual vector $\hat{\varepsilon}$.

The predicted values for the following 6 months, together with the data and the estimated mean value function, are in Figure 5.3.4.

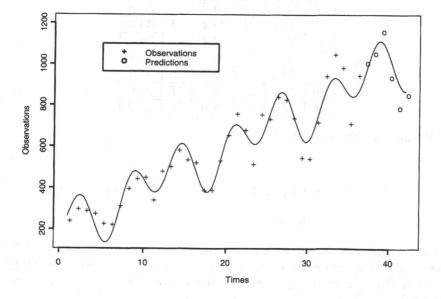

Figure 5.3.4.

These predicted values are given in the following table:

t	37	38	39	40	41	42
$\hat{x}(t)$	1004	1050	1158	934	785	849

The values $m_{\hat{\gamma}}(t); t = 37, 38, ..., 42$, of the estimated mean value function, which can be regarded as predictors for $x(t); t = 37, 38, ..., 42$ under the assumption that $\varepsilon(.)$ is a white noise, are given in the next table:

t	37	38	39	40	41	42
$m_{\hat{\gamma}}(t)$	973	1061	1050	948	855	872

Example 5.3.3. The following data are weekly observed prices, in Slovak crowns, of gasoline in Slovakia during the first forty 48 weeks of the year 2000.

t	$x(t)$	t	$x(t)$	t	$x(t)$
1	31.6	17	32.4	33	31.8
2	30.5	18	31.9	34	32.2
3	30.7	19	32.2	35	32.6
4	31.6	20	32.8	36	32.6
5	31.6	21	33.7	37	33.1
6	32.0	22	34.2	38	34.0
7	32.1	23	34.2	39	34.6
8	32.1	24	34.6	40	34.3
9	32.1	25	34.4	41	34.3
10	32.6	26	34.2	42	33.8
11	33.7	27	34.2	43	33.2
12	33.7	28	34.2	44	34.2
13	33.5	29	34.6	45	35.1
14	33.5	30	33.9	46	35.4
15	33.5	31	33.5	47	34.0
16	32.4	32	32.6	48	33.3

We use for these data the NRM

$$X(t) = \beta_1 + \beta_2 t + \sum_{i=1}^{k} \left(\beta_i^1 \cos \lambda_i t + \beta_i^2 \sin \lambda_i t \right) + \varepsilon(t)$$
$$= m_{\gamma}(t) + \varepsilon(t); t = 1, 2, ...,$$

with an unknown k and with an unknown regression parameter γ. We compute, using the observed vector $x = (x(1), x(2), ..., x(48))'$, the OLSE $\tilde{\beta} = (\tilde{\beta}_1, \tilde{\beta}_2)'$ of $\beta = (\beta_1, \beta_2)'$. Let $\tilde{\varepsilon}$ be the vector of residuals with components

$$\tilde{\varepsilon}(t) = x(t) - \tilde{\beta}_1 - \tilde{\beta}_2 t; t = 1, 2, ..., 36.$$

The periodogram of the residuals $\tilde{\varepsilon}$ was computed again and it can be seen from this periodogram that $k = 3$ and that the frequences $\lambda_1 = 0.125\pi$, $\lambda_2 = 0.25\pi$, and $\lambda_3 = 0.041\pi$ are significant. These frequencies correspond to periodicities $T_1 = 16$, $T_2 = 8$, and $T_3 = 48$ weeks. We can see this periodogram in Figure 5.3.5.

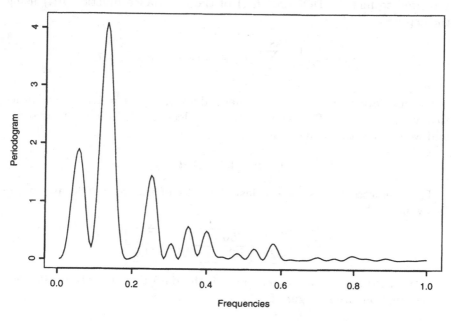

Figure 5.3.5.

Then again the ordinary least squares method was used to find the OLSE $\hat{\gamma}$. We get

$$\hat{\beta}_1 = 31.9, \hat{\beta}_2 = 0.51,$$

and

$$\hat{\lambda}_1 = 0.12\pi, \hat{\beta}_1^1 = -0.46, \hat{\beta}_2^1 = -0.68,$$
$$\hat{\lambda}_2 = 0.25\pi, \hat{\beta}_2^1 = -0.03, \hat{\beta}_2^2 = -0.50,$$
$$\hat{\lambda}_3 = 0.04\pi, \hat{\beta}_3^1 = -0.44, \hat{\beta}_3^2 = 0.25,$$

The estimated mean value function is

$$m_{\hat{\gamma}}(t) = 31.9 + 0,51t$$
$$-0.46\cos 0.12\pi t - 0.68\sin 0.12\pi t$$
$$-0.03\cos 0.25\pi t - 0.50\sin 0.25\pi t$$
$$-0.44\cos 0.04\pi t + 0.25\sin 0.04\pi t.$$

Using this mean value function, the vector of the ordinary least squares residuals $\hat{\varepsilon}$ with components

$$\hat{\varepsilon}(t) = x(t) - m_{\hat{\gamma}}(t); t = 1, 2, ..., 36,$$

was used to find the DOOLSE $\hat{R}(.)$ of the covariance function $R(.)$ using the expression

$$\hat{R}(t) = \frac{1}{48-t} \sum_{s=1}^{48-t} \hat{\varepsilon}(s+t)\hat{\varepsilon}(s); t = 0, 1, ..., 47.$$

For our data $\hat{R}(0) = 0.17$ and other values of $\hat{R}(t); t = 1, 2, ...,$ are practically equal to zero. Thus $\varepsilon(.)$ can be regarded as a white noise time series and we can use the predictors

$$\hat{x}(48 + d) = m_{\hat{\gamma}}(48 + d); d = 1, 2, ...$$

The predicted values for the last 4 weeks of the year 2000 are in the following table

t	49	50	51	52
$\hat{x}(t)$	32.9	32.8	33.1	33.6

These predicted values, together with the data and the estimated mean value function, are in Figure 5.3.6.

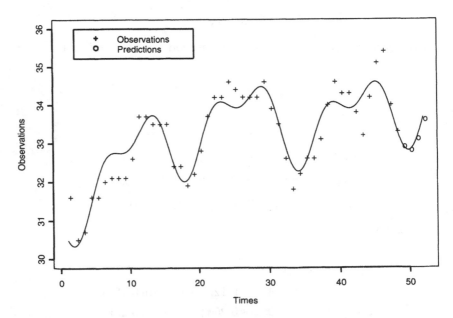

Figure 5.3.6.

Often we can consider monthly data during q years as observations of *a nonlinear multivariate regression model with seasonal components*. Such a model is given by

$$X = (I_3 \otimes F)B + C_\vartheta + \varepsilon = M_\gamma + \varepsilon; \gamma = (B, \vartheta),$$

where X is the $12 \times q$ random matrix, $X = (X_1 X_2 ... X_q)$, with columns $X_i = (X_i(1), X_i(2), ..., X_i(12))'; i = 1, 2, ..., q$, which are monthly observations in q years, F is the 12×2 design matrix of the linear trend,

$$F = \begin{pmatrix} 1 & 1 & . & . & . & 1 \\ 1 & 2 & . & . & . & 12 \end{pmatrix}',$$

where $B = (\beta_1 \beta_2 ... \beta_q)$ with $\beta_i = (\beta_{i,1}, \beta_{i,2})'$ are parameters of the linear trends of the time series $X_i(.)$ at the ith year, $i = 1, 2, ..., q$, and the $12 \times q$ matrix C_ϑ has columns $C_{i,\vartheta}; i = 1, 2, ..., q$, with

$$C_{i,\vartheta}(t) = \sum_{i=1}^{k} (\gamma_i^1 \cos \lambda_i t + \gamma_i^2 \sin \lambda_i t); t = s + (i-1)12,$$

where $s = 1, 2, ..., 12$ and $i = 1, 2, ..., q$.
The function

$$C_\vartheta(t) = \sum_{i=1}^{k} (\gamma_i^1 \cos \lambda_i t + \gamma_1^2 \sin \lambda_i t); t = 1, 2, ...,$$

with parameter $\vartheta = (\lambda', \gamma')'$ describes the seasonal components of the observed time series $X(.)$.
For predicted random variables $X_{q+1}(t); t = 1, 2, ..., 12$, we assume that

$$E[X_{q+1}(t)] = \beta_{q+1,1} + \beta_{q+1,2}(t) + C_\vartheta(12q + t); t = 1, 2, ..., 12.$$

As predictors $\tilde{X}_{q+1}(t)$ of $X_{q+1}(t); t = 1, 2, ..., 12$, we take

$$\tilde{X}_{q+1}(t) = \tilde{\beta}_{q+1,1} + \tilde{\beta}_{q+1,2}t + C_{\tilde{\vartheta}}(12q + t) + \tilde{r}_t' \tilde{\Sigma}^{-1}(Vec(X) - vec(M_{\tilde{\gamma}})),$$

where $\tilde{\beta}_{q+1}, \tilde{\vartheta}$, and $\tilde{\gamma}$ are estimators of the unknown parameters β_{q+1} and ϑ, and $\tilde{r}_t$ and $\tilde{\Sigma}$ are some estimators of $r_t = Cov(X_{q+1}(t); Vec(X))$ and $\Sigma = Cov(Vec(X))$.

Example 5.3.4. The following data are the monthly observed numbers of customers $x_i(t); t = 1, 2, ..., 12; i = 1, 2, 3$, of new products in 3 years:

t	$x_1(t)$	$x_2(t)$	$x_3(t)$
1	4.5	3.9	2.5
2	6.0	4.0	3.0
3	8.9	6.5	4.9
4	11.9	9.1	7.0
5	13.5	10.4	7.6
6	13.8	10.0	6.9
7	14.4	10.2	7.0
8	16.8	11.0	8.9
9	20.9	14.6	9.4
10	25.0	19.0	14.1
11	27.6	21.0	15.8
12	28.2	21.1	14.9

For these data we use the model

$$
X_i(t) = \beta_{i,1} + \beta_{i,2}t
$$
$$
+ \sum_{i=1}^{k} \left(\gamma_i^1 \cos \lambda_i[(i-1)12 + t] + \gamma_i^2 \sin \lambda_i[(i-1)12 + t] \right) + \varepsilon_i(t),
$$

where $t = 1, 2, ..., 12$ and $i = 1, 2, 3$. The estimates of the unknown parameters of this model we find as follows. Using X_i we find the OLSEs $\hat{\beta}_i$ for β_i for $i = 1, 2, 3$. We get

$$
\begin{aligned}
\hat{\beta}_1 &= (1.83, 2.18)', \\
\hat{\beta}_2 &= (1.34, 1.61)', \\
\hat{\beta}_3 &= (1.02, 1.15)'.
\end{aligned}
$$

Then we use the double linear trend prediction method described in Example 4.3.1 to get the estimate $\tilde{\beta}_4$ of β_4 and we get

$$
\tilde{\beta}_4 = (0.58, 0.62)'.
$$

Then we can compute the periodogram of the residuals $Vec(X - (I \otimes F)\hat{B})$, where $\hat{B} = (\hat{\beta}_1 \hat{\beta}_2 \hat{\beta}_3)$. This periodogram is in Figure 5.3.7. and we can see that there are two significant frequencies, $\tilde{\lambda}_1 = 0.33\pi$ and $\tilde{\lambda}_2 = 0.16\pi$.

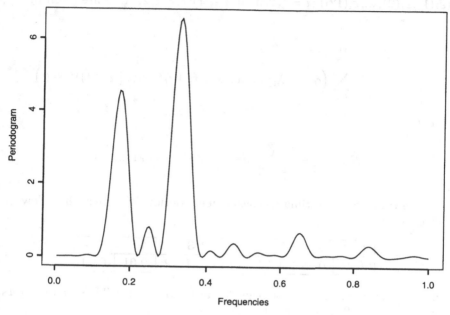

Figure 5.3.7.

In the LRM

$$\hat{X}_i(t) = \hat{\beta}_{i,1} + \hat{\beta}_{i,2}t$$
$$+ \sum_{i=1}^{2} \gamma_i^1 \cos \tilde{\lambda}_i [(i-1)12 + t] + \gamma_i^2 \sin \tilde{\lambda}_i [(i-1)12 + t] + \eta_i(t)$$

with unknown parameters $\gamma_i = (\gamma_i^1, \gamma_i^2)'; i = 1, 2$, where $t = 1, 2, ..., 12$ and $i = 1, 2, 3$, we can compute the OLSEs $\hat{\gamma}_i$ of γ_i for $i = 1, 2$. These are

$$\hat{\gamma}_1 = (-1.20, -1.24)$$
$$\hat{\gamma}_2 = (0.85, 0.62).$$

Let $Vec(\hat{\varepsilon}) = (\hat{\varepsilon}_1'\hat{\varepsilon}_2'\hat{\varepsilon}_3')$ be the 36×1 vector of residuals, where $\hat{\varepsilon}_i = (\hat{\varepsilon}_i(1), \hat{\varepsilon}_i(2), ..., \hat{\varepsilon}_i(12))'; i = 1, 2, 3,$ and $\hat{\varepsilon}_i(t); t = 1, 2, ..., 12$ are given by

$$\hat{\varepsilon}_i(t) = X_i(t) - \hat{\beta}_{i,1} - \hat{\beta}_{i,2}t$$
$$- \sum_{i=1}^{2} \left(\hat{\gamma}_i^1 \cos \tilde{\lambda}_i[(i-1)12 + t] + \hat{\gamma}_i^2 \sin \tilde{\lambda}_i[(i-1)12 + t] \right).$$

Let

$$\hat{R}(t) = \frac{1}{36 - t} \sum_{s=1}^{36-t} \hat{\varepsilon}(s+t)\hat{\varepsilon}(s); t = 0, 1, ..., 35.$$

Some values of this estimated covariance function $\hat{R}(.)$ are in the following table:

t	0	1	2	3	4	5
$\hat{R}(t)$	0.35	-0.04	-0.12	-0.12	0.01	0.09

and we can see that $\hat{\varepsilon}(.)$ can be regarded as a white noise. Thus as predictors for $X_4(t); t = 1, 2, ..., 12$, we take

$$\tilde{X}_4(t) = \tilde{\beta}_{4,1} + \tilde{\beta}_{4,2}t$$
$$+ \sum_{i=1}^{2} \left(\hat{\gamma}_i^1 \cos \tilde{\lambda}_i(36 + t) + \hat{\gamma}_i^2 \sin \tilde{\lambda}_i(36 + t) \right)$$
$$= 0.58 + 0.62t$$
$$-1.20 \cos 0.33\pi(36 + t) - 1.24 \sin 0.33\pi(36 + t)$$
$$+0.85 \cos 0.16\pi(36 + t) + 0.62 \sin 0.16\pi(36 + t).$$

Predicted values for the fourth year are in the following table:

t	1	2	3	4	5	6
$\tilde{x}_4(t)$	0.1	2.5	4.0	5.8	4.8	3.1

t	7	8	9	10	11	12
$\tilde{x}_4(t)$	2.6	4.0	6.1	7.4	7.2	6.8

The data, the predicted values, together with the estimated mean value function, are in Figure 5.3.8.

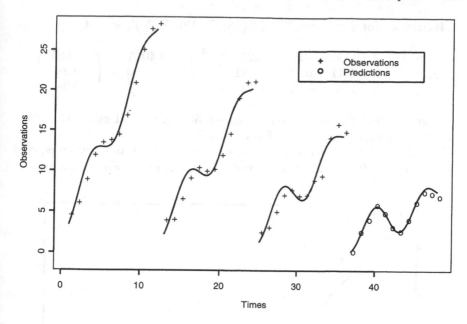

Figure 5.3.8.

Example 5.3.5. In the following table there are simulated data from the model

$$X(t) = \gamma_1 \exp(-\gamma_2 t) + \varepsilon(t); t = 1, 2, ..., 20,$$

where $\gamma = (\gamma_1, \gamma_2)' = (10, -0.1)'$. We asume that $\varepsilon(.)$ is an AR(3) time series with parameter $a = (0.1, 0.3, -0.6)'$ and with the variance $\sigma^2 = 0.5$ of a white noise.

t	1	2	3	4	5	6	7	8	9	10
$x(t)$	9.4	8.4	7.8	7.9	7.4	5.7	4.79	3.4	4.1	2.7
t	11	12	13	14	15	16	17	18	19	20
$x(t)$	4.6	2.9	3.9	1.6	2.0	1.6	0.8	2.8	1.2	3.1

The OLSE $\hat{\gamma}$ of γ is $\hat{\gamma} = (10.6, -0.1)'$. Some values of the estimated covariance function $\hat{R}(.)$ are in the following table:

t	0	1	2	3	4	5
$\hat{R}(t)$	0.71	-0.11	0.25	-0.40	-0.22	-0.10

We use empirical predictors given by

$$\tilde{X}(20 + d) = \hat{\gamma}_1 \exp(-\hat{\gamma}_2(20 + d)) + \hat{r}'_d \hat{\Sigma}^{-1}(X - m_{\hat{\gamma}}); d = 1, 2, 3,$$

where $X = (X(1), ..., X(20))', \hat{\Sigma}$ is the DOOLSE of the covariance matrix $\Sigma, \hat{r}_d$ is the DOOLSE of the vector $r_d = Cov(X; X(20 + d))$, and $m_{\hat{\gamma}} = E_{\hat{\gamma}}[X]$.

Remark. For $d = 1$ and $X = (X(18), X(19), X(20)'$, we get

$$\hat{\Sigma}^{-1}\hat{r}_1 = \begin{pmatrix} 0.71 & -0.11 & 0.25 \\ -0.11 & 0.71 & -0.11 \\ 0.25 & -0.11 & 0.71 \end{pmatrix}^{-1} \begin{pmatrix} -0.40 \\ 0.25 \\ -0.11 \end{pmatrix} = \begin{pmatrix} -0.55 \\ 0.28 \\ 0.08 \end{pmatrix}$$

and we see that the vector $(0.08, 0.26, -0.55)'$ is a good estimator of the vector $a = (0.1, 0.3, -0.6)'$, see Example 4.2.7.

The data, together with the estimated mean value and with predicted values $\tilde{x}(21) = 0.67, \tilde{x}(22) = 1.45$, and $\tilde{x}(23) = 0$, are in Figure 5.3.9.

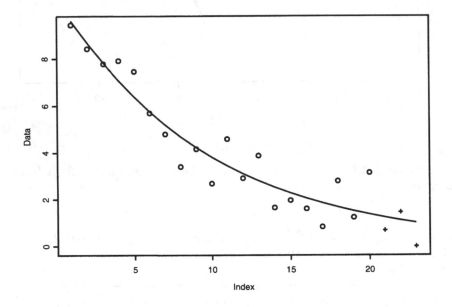

Figure 5.3.9.

References

Abraham, B. and Ledolter, J. (1983). *Statistical Methods for Forecasting.* Wiley, New York.

Anderson, T.W. (1971). *The Statistical Analysis of Time Series.* Wiley, New York.

Anderson, T.W. (1984). *An Introduction to Multivariate Statistical Analysis.* 2nd ed. Wiley, New York.

Andĕl, J. (1976). *Statistická analýza časových řad.* SNTL, Praha.

Arnold, S.F. (1981). *The Theory of Linear Models and Multivariate Analysis.* Wiley, New York.

Azais, J.M., Bardin, A., and Dhorne, T. (1993). MINQUE, maximum likelihood estimation and Fisher Scoring Algorithm for nonlinear variance models. *Statistics,* **24,** 205-213.

Beach, C.M. and MacKinnon, J.G. (1978). A maximum likelihood procedure for regression with auto-correlated errors. *Econometrica,* **46,** 51-58.

Bhansali, R.J. (1971). Effect of not knowing the order of autoregression on the mean squared error of prediction. *J. Amer. Statist. Assoc.,* 588-597.

Box, M.J. (1971). Bias in nonlinear estimation. *J. Roy. Statist. Soc.,* **33,** 171-201.

Box, G.E.P. and Jenkins, G.M. (1976). *Time Series Analysis, Forecasting and Control,* rev. ed. Holden Day, San Francisco.

Brockwell, P.J. and Davis, R.A. (1987). *Time Series: Theory and Methods.* Springer-Verlag, New York.

Brockwell, P.J. and Davis, R.A. (1996). *Introduction to Time Series and Forecasting.* Springer-Verlag, New York.

Cook, R.D. and Tsai, C.L. (1985). Residuals in nonlinear regression. *Biometrika,* **72,** 23-29.

Christensen, R. (1987). *Plane Answers to Complex Questions: The Theory of Linear Models*. Springer-Verlag, New York.

Christensen, R. (1990). The equivalence of prediction from universal kriging and intrinsic random function kriging. *Math. Geology*, **22**, 655 - 664.

Christensen, R. (1991). *Linear Models for Multivariate, Time Series and Spatial Data*. Springer-Verlag, New York.

Cressie, N. (1988). Kriging nonstationary data. *J. Amer. Statist. Assoc.*, **81**, 625-634.

David, M. (1977). *Geostatistical Ore Reserve Estimations*. Elsevier, New York.

Diggle, P.J. (1996). *Time Series: a Biostatistical Introduction*. Clarendom Press, Oxford.

Eaton, M.L. (1985) The Gauss-Markov theorem in multivariate analysis. In: *Multivariate Analysis VI*, edited by P.R. Krishnaiah. North-Holland, Amsterdam.

Fuller, W.A. (1976). *Introduction to Statistical Time Series*. Wiley, New York.

Fuller, W.A. and Hasza, D.P. (1981). Properties of predictors for autoregressive time series. *J. Amer. Statist. Assoc.*, **76**, 155-161.

Gallant, A.R. (1987). *Nonlinear Statistical Modeling*. Wiley, New York.

Gilchrist, W.G. (1976). *Statistical Forecasting*. Wiley, Chichester.

Goldberger, A.S. (1962). Best linear unbiased prediction in the generalized linear regression model. *J. Amer. Statist. Assoc.*, **57**, 369-375.

Gumpertz, M.L. and S. G. Pantula, S.G.(1992). Nonlinear regression with variance components. *J. Amer. Statist. Assoc.*, **87**, 201-209.

Harvey, A.C. (1990). *The Econometric Analysis of Time Series*. MIT Press, Cambridge,MA.

Harvey, A.C. (1994). *Time Series Models*. MIT Press, Cambridge, MA.

Harvey, A.C. (1996). *Forecasting, Structural Time Series and the Kalman Filter*. Cambridge University Press, Cambridge, UK.

Harville, D.A. (1977). Maximum likelihood approach to variance components estimation and related problems. *J. Amer. Statist. Assoc.*, **72**, 320-340.

Harville, D.A. (1985). Decomposition of prediction error. *J. Amer. Statist. Assoc.*, **80**, 132-138.

Harville, D.A. (1990). BLUP and beyond. In: *Advances in Statistical Methods for Improvement of Livestock.* Springer-Verlag, New York.

Harville, D.A and Jeske, D.R. (1992). Mean squared error of estimation or prediction under a general linear model. *J. Amer. Statist. Assoc.*, **87**, 724-731.

Jennrich, R.I. (1969). Asymptotic properties of nonlinear least squares estimators. *Ann. Math. Statist.*, **40**, 633-643.

Journel, A.G. (1977). Kriging in terms of predictions. *J. Internat. Assoc. Math. Geol.*, **9**, 563-586.

Journel, A.G. and Hüijbregts, Ch.J. (1978). *Mining Geostatistics.* Academic Press, London.

Kackar, R.N. and Harville, D.A. (1984). Approximation for standard errors of estimation in mixed linear models. *J. Amer. Statist. Assoc.*, **79**, 853-862.

Kubáček, L. (1988). *Foundations of Estimation Theory.* Elsevier, Amsterdam.

Kubáčková, L., Kubáček, L., and Kukuča, J. (1988). *Probability and Statistics in Geodesy and Geophysics.* Elsevier, Amsterdam.

Mardia, K.V. and Marshall, R.J. (1984). Maximum likelihood estimation of models for residual covariance in spatial regression. *Biometrika*, **71**, 135-146.

McGilchrist, C.A. (1994). Estimation in generalized mixed models. *J. Roy. Statist. Soc. B*, **56**, 61-69.

Parzen, E. (1961). An approach to time series analysis. *Ann. Math. Statist.*, **32**, 951-959.

Parzen, E. (1967). *Time Series Analysis Papers.* Holden-Day, London.

Pázman, A. (1993). *Nonlinear Statistical Models.* Kluwer Academic, Dordrecht.

Prasad, N.G.N. and Rao, J.N.K. (1990). The estimation of the mean squared error of small-area estimation. *J. Amer. Statist. Assoc.*, **85**, 163-171.

Priestley, M.B. (1981). *Spectral Analysis and Time Series.* Academic Press, London.

Rao, C.R. and Kleffe, J. (1998). *Estimation of Variance Components and Applications.* North- Holland, Amsterdam.

Rattkowsky, D.A. (1983). *Nonlinear Regression Modeling.* Marcel Dekker, New York.

Ripley, B.D. (1981). *Spatial Statistics.* Wiley, New York.

Seber, G.A.F. (1977). *Regression Analysis.* Wiley, New York.

Seely, J. (1971). Linear spaces and unbiased estimation-Application to mixed linear model. *Ann. Math. Statist.*, **41**, 1725-1734.

Stein, M.L. (1999). *Interpolation of Spatial Data. Some Theory for Kriging.* Springer-Verlag, New York.

Štulajter, F. (1989). *Estimation in Random Processes* (in Slovak). Alfa, Bratislava.

Štulajter, F. and Hudáková, J. (1991a). An approximate least squares estimator in nonlinear regression and some of its properties. *Proc. Probastat'91*, Bratislava.

Štulajter, F. (1991b). Cosistency of linear and quadratic least squares estimators with covariance stationary errors. *Appl. Math.*, **36**, 149-155.

Štulajter, F. (1992a). Mean squared error matrix of an approximate LSE in a nonlinear regression model with correlated errors. *Acta Math. Univ. Comenian.*, **LXI**, 251-261.

Štulajter, F. (1992b). Some aspects of kriging. *Transactions of the 11th Prague Confer.*, Prague, pp. 397-402.

Štulajter, F. (1994a). On estimation of a covariance function of stationary errors in a nonlinear regression model. *Acta Math. Univ. Comenian.*, **LXIII**, 107-116.

Štulajter, F. (1994b). Mean squared error of prediction by kriging in linear models with AR(1) errors. *Acta Math. Univ. Comenian.*, **LXIII** 2, 247-254.

Štulajter, F. and Stano, S. (1996). A comparison of some predictors in nonlinear regression models. *Tatra Mount. Math. Publ.*, **7**, 305-309.

Štulajter, F. (1997). Predictions in nonlinear regression models. *Acta Math. Univ. Comenian.*, **LXVI**, 71-81.

Štulajter, F. (1999). Prediction in multivariate regression models. *Tatra Mount. Math. Publ.*, **17**, 265-272.

Štulajter, F. (2001). Predictions in time series using multivariate regression models. *J. Time Ser. Anal.*, **22**, 365-373.

Sweeting, T.J. (1980). Uniform asymptotic normality of the maximum likelihood estimation. *Ann. Statist.*, **8**, 1375-1381.

Taylor, S.J. (1986). *Modeling Financial Time Series*. Wiley, Chichester, UK.

Toyooka, Y. (1982). Prediction error in a linear model with estimated parameters. *Biometrika*, **69**, 453-459.

Usami, Y. and Huzii, M. (1993). Estimation of time series regression with a nonstationary error process. *J. Time Ser. Anal.*, **96**, 105-118.

Witkovský, V. (1996). On variance-covariance component estimation in linear models with AR(1) disturbances. *Acta Math. Univ. Comenian.*, **LXV**, 123-139.

Wu, C.F. (1981). Asymptottic theory of nonlinear least squares estimators. *Ann. Statist.*, **9**, 5o1-513.

Zimmerman, D.L. and Cressie, N. (1992). Mean squared error in spatial linear model with estimated regression parameters. *Ann. Inst. Math. Statist.*, **44**, 27-43

Index

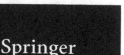